SPRINGER TRACTS IN MODERN PHYSICS

Ergebnisse
der exakten Natur-
wissenschaften

Volume 66

Springer-Verlag Berlin Heidelberg GmbH 1973

Manuscripts for publication should be adressed to:

G. Höhler, Institut für Theoretische Kernphysik der Universität, 75 Karlsruhe 1, Postfach 6380

Proofs and all correspondence concerning papers in the process of publication should be addressed to:

E. A. Niekisch, Kernforschungsanlage Jülich, Institut für Technische Physik, 517 Jülich, Postfach 365

ISBN 978-3-662-39407-6 ISBN 978-3-662-40468-3 (eBook)
DOI 10.1007/978-3-662-40468-3

 Library of Congress Catalog Card Number 25-9130.
Originally published by Springer-Verlag Berlin Heidelberg New York in 1973.
Softcover reprint of the hardcover 1st edition 1973

Quantum Statistics in Optics and Solid-State Physics

Contents

Statistical Theory of Instabilities in Stationary Non-equilibrium Systems with Applications to Lasers and Nonlinear Optics

R. GRAHAM

Contents

A. General Part

1. Introduction and General Survey

The transition of a macroscopic system from a disordered, chaotic state to an ordered more regular state is a very general phenomenon as is testified by the abundance of highly ordered macroscopic systems in nature. These transitions are of special interest, if the change in order is structural, i.e. connected with a change in the symmetry of the system's state.

The existence of such symmetry changing transitions raises two general theoretical questions. In the first place one wants to know the conditions under which the transitions occur. Secondly, the mechanisms which characterize them are of interest.

Since the entropy of a system decreases, when its order is increased, it is clear from the second law of thermodynamics that transitions to states with higher ordering can only take place in open systems interacting with their environment.

Two types of open systems are particularly simple. First, there are systems which are in thermal equilibrium with a large reservoir prescribing certain values for the intensive thermodynamic variables. Structural changes of order in such systems take place as a consequence of an instability of all states with a certain given symmetry. They are known as second order phase transitions. Both the possibility of their occurrence and their general mechanisms have been the subject of detailed studies for a long time.

A second, simple class of open systems is formed by stationary nonequilibrium systems. They are in contact with several reservoirs, which are not in equilibrium among themselves.

These reservoirs impose external forces and fluxes on the system and thus prevent it from reaching an equilibrium state. They rather keep it in a nonequilibrium state, which is stationary, if the properties of the various reservoirs are time independent.

Structural changes of order in such systems again take place, if all states with a given symmetry become unstable. They were much less investigated in the past, and moved into the focus of interest only recently, although they occur quite frequently and give, in fact, the only clue to the problem of the self-organization of matter. The general conditions under which such instabilities occur where investigated by Glansdorff and Prigogine in recent publications [1 – 4]. A statistical foundation of their theory was recently given by Schlögl [5]. The general picture, emerging from the results in [1 – 4] may be summarized for our purposes as follows (cf. Fig. 1):

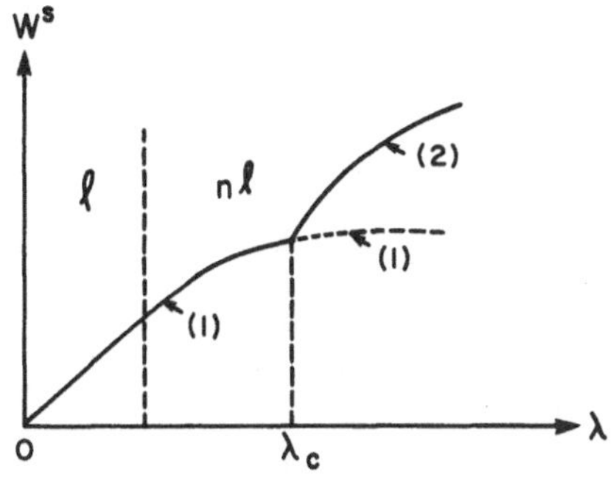

Fig. 1. Two branches of stationary nonequilibrium states connected by an instability (see text)

Starting with a system in a stable thermal equilibrium state (point 0 in Fig. 1), one may create a branch of stationary nonequilibrium states by applying an external force λ of increasing strength. If λ is sufficiently small one may linearize the relevant equations of motion with respect to the small deviations from equilibrium (region *l* in Fig. 1). In this region one finds that all stationary nonequilibrium states are stable if the thermal equilibrium state is stable. If λ becomes sufficiently large, the linearization is no longer valid (region *nl* in Fig. 1). In this case, it is possible that the branch (1) becomes unstable (dotted line in Fig. 1) for $\lambda > \lambda_c$, where λ_c is some critical value, and a new branch (2) of states is followed by the system. This instability may lead to a change of the symmetry of the stable states. Assume that the states on branch (2) have a lower symmetry (i.e. higher order) than the states on branch (1). Since for $\lambda = \lambda_c$ the lower symmetry of branch (2) degenerates to the higher symmetry of branch (1), the states of branch (2) merge continuously with the states of branch (1).

A simple example is shown in Fig. 2. There, the system is viewed as a particle moving with friction in a potential $\phi^s(w)$ with inversion symmetry $\phi^s(w) = \phi^s(-w)$. The external force λ is assumed to deform the potential without changing its symmetry. Three typical shapes for $\lambda \gtreqless \lambda_c$ are shown. The stationary states w^s, given by the minima of the potential, are plotted as a function of λ (broad line). For $\lambda = \lambda_c$ the branch (1) of stationary states having inversion symmetry becomes unstable and a new branch (2) of states, lacking inversion symmetry, is stable.

There are many physically different systems, which show this general behaviour. A well known hydrodynamical example is furnished by the convective instability of a liquid layer heated from below (Benard instability). The spatial translation invariance in the liquid layer at rest is

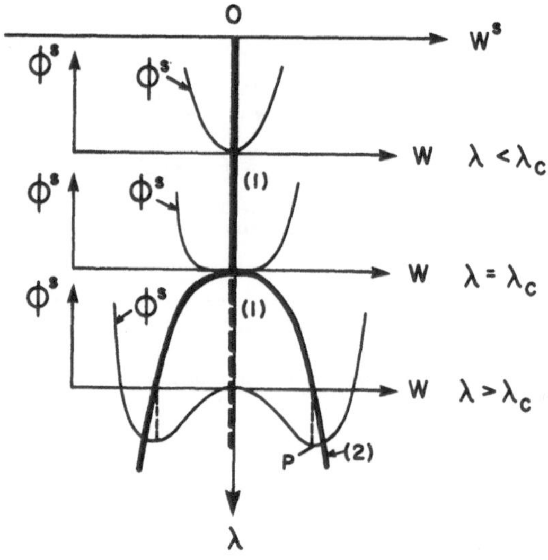

Fig. 2. Stationary state w^s (thick line) of a particle moving with friction in a potential $\phi^s(w)$ with inversion symmetry, plotted as a function of an external force λ

broken by the formation of a regular lattice of convection cells in the convective state (cf. [4, 6]). Other examples discussed in the literature are periodic oscillations of concentrations of certain substances in autocatalytic reactions [4, 7] which also occur in biological systems, or periodic features in the dynamics of even more complex systems [4] (e.g. Volterra cycles).

While the Glansdorff-Prigogine theory predicts the occurrence of the instabilities, so far little work has been concerned with the general mechanisms of the transitions. In the present paper we want to address ourselves to this question. As in the case of phase transitions, the general mechanisms can best be analyzed by looking at the fluctuations near the basic instability, which were neglected completely so far. This is the subject of the first half (part A) of this paper.

Experimentally, the fluctuations near the instabilities in the systems mentioned above have not yet been determined, although, in some cases (hydrodynamics) experiments seem to be possible and would be very interesting, indeed. Fortunately, however, a whole new class of instabilities has been discovered in optics within the last ten years, for which the fluctuations are more directly measurable than in the cases mentioned above. These are the instabilities which give rise to laser action [8] and induced light emission by the various scattering processes of nonlinear optics [9]. The fluctuations in optics are connected with the emitted light and can, hence, be measured directly by photon counting methods [10].

More indirect methods like light scattering would have to be used in other cases. In part B the considerations of part A are applied to a number of optical instabilities.

In order to put the optical instabilities into the general scheme outlined in Fig. 1, we look at a simple example. Let us consider an optical device, in which a stimulated scattering process takes place between the mirrors of a Perot Fabry cavity, which emits light in a single mode pattern. An example would be a single mode laser or any other optical oscillator, like a Raman Stokes oscillator or a parametric oscillator. A diagram like Fig. 1 is obtained by plotting (besides other variables) the real part of the complex mode amplitude β versus the pump strength λ, which is proportional to the intensity of the pumping source (Fig. 3a). Neglecting all fluctuations (as we did in Fig. 1), the simple theory of such devices [11] gives the following general behavior.

For very weak pumping the system may be described by equations, which are linearized with respect to the deviations from thermal equi-

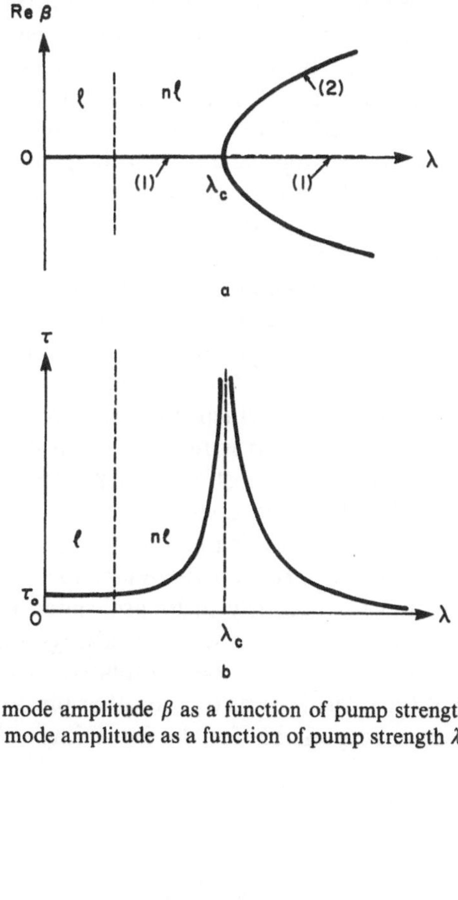

Fig. 3a. Real part of mode amplitude β as a function of pump strength λ (see text)
b. Relaxation time of mode amplitude as a function of pump strength λ

librium. The result for the amplitude of the oscillator mode is zero. Furthermore, one obtains some finite, constant value for the relaxation time τ of the amplitude, which is plotted schematically in Fig. 3b. No instability, whatsoever, is possible in this linear domain, in agreement with the general result.

With increased pumping, the nonlinearity of the interaction of light and matter has to be taken into account by linearizing around the stationary state, rather than around thermal equilibrium. The stationary solution for the complex amplitude of the oscillator mode is still zero. The deviations from thermal equilibrium are described by some other variables, which are not plotted in Fig. 3a (e.g. the occupation numbers of the atomic energy levels in the laser case). In contrast to the case of very weak pumping, the relaxation time of the mode amplitude now goes to infinity for some pumping strength $\lambda = \lambda_c$ indicating the onset of instability of this mode. For $\lambda > \lambda_c$ a new branch of states is found to be stable with non-zero mode amplitude and a finite relaxation time τ. The zero-amplitude branch is unstable.

The two different branches of states have different symmetries. All states on the zero-amplitude branch have a complete phase angle rotation invariance. The phase symmetry is broken on the finite-amplitude branch, since the complex mode amplitude has a fixed, though arbitrary, phase on this branch. The broken symmetry implies the existence of a long range order in space and (or) time. It should be noted, however, that this result is modified if fluctuations are taken into account. In summary, we find complete agreement with the general behaviour, outlined in Fig. 1. In particular, the importance of the nonlinear interaction between light and matter is clearly born out.

It is instructive to compare this phenomenological picture with the microscopic picture of the same instability. From the microscopic point of view the region *l* is the region where fluctuation processes alone are important (spontaneous emission). In the region *nl* stimulated emission becomes important. In fact, it is the same nonlinearity in the interaction of light and matter which gives rise to stimulated emission and the instability. The threshold is reached when it is more likely that a photon stimulates the emission of another photon, rather than if the photon is dissipated by other processes.

This picture of the instability is much more general than the optical example, from which it was derived here. In fact, in as much as all macroscopic instabilities have necessarily to be associated with boson modes because of their collective nature, we may always interpret the onset of instability as a taking over of the stimulated boson emission over the annihilation of the same bosons due to other processes. The stimulated emission process, responsible for the instability in this microscopic picture, is due to the nonlinearity, which was found by Glansdorff and Prigogine to be necessary for the onset of instability.

If the threshold of instability is passed, the number of bosons grows until a saturation effect due to induced absorption determines a final stationary state. In this state the coherent induced emission and re-absorption of bosons constitutes a long range order in space and (or) time.

The degree to which this order is modified by fluctuations depends on the spatial dimensions of the system. For systems with short range interactions there exists no order of infinite range in less than two spatial dimensions [12]. Broken symmetries and long range order are found in such systems only if fluctuations are neglected. If the latter are included, the symmetry is always restored by a diffusion of the parameter, which characterizes the symmetry in question (the phase angle in the above example). This slow phase diffusion is a well known phenomenon for the single mode oscillator discussed before (cf. [8]). The same phenomenon is found in all optical examples, which are discussed in part B.

Therefore, symmetry considerations also play an important role for those instabilities in which symmetry changes are finally restored by fluctuations. Furthermore, the fluctuations are frequently very weak and need a long time or distance to restore the full symmetry. Therefore, we find it useful to consider all these instabilities together from the common point of view, that they change the symmetry of the stationary state without fluctuations. They are called "symmetry changing transitions" in the following.

We now give a brief outline of the material in this article. The paper is divided into two parts. The first part A is devoted to a general phenomenological theory of fluctuations in the vicinity of a symmetry changing instability. In the second part B the general results of part A are applied to a number of examples from laser physics and nonlinear optics. Throughout the whole paper we restrict ourselves to systems which are stationary, Markoffian and continuous. These basic assumptions are introduced in section 2.1. The fundamental equations of motion can then be formulated along well known lines either as a Fokker-Planck equation (cf. 2.1.a) or as a set of Langevin equations (cf. 2.1.b). In this frame, the phenomenological quantities, which describe the system's motion are a set of drift and diffusion coefficients. They depend on the system's variables and a set of time independent parameters, which describe the external forces, acting on the system. All other quantities can, in principle, be derived from the drift and diffusion coefficients. However, in many cases it is preferable to use the stationary probability distribution as a phenomenological quantity, which is given, rather than derived from the drift and diffusion coefficients. This is a very common procedure in equilibrium theory, where the stationary distribution is always assumed to be known and taken to be the canonical distribution. For stationary nonequilibrium problems this procedure is unusual, although, as will be shown, it can have many advantages. It is an important part of our phenomenological approach. If the stationary distribution is known, it can be used to re-express the drift coefficients in a general way (cf. 2.2), which is a direct generalization of the familiar linear relations between fluxes and forces in irreversible thermodynamics [13], valid near equilibrium states.

The formal connection with equilibrium theory is investigated further by generalizing the Onsager Machlup formulation of linear irreversible thermodynamics [14 – 16] to include also the nonlinear theory of stationary states far from equilibrium (cf. 2.3).

Since the knowledge of the stationary distribution is the starting point of our phenomenological theory, section 3 is devoted to a detailed study of its general properties. Special attention is paid to the relations between the theory which neglects fluctuations and the theory which includes fluctuations.

In 3.1, we show, that without fluctuations, the system may be in a variety of different stable stationary states, whereas the inclusion of fluctuations leads to a unique and stable distribution over these states. This result is used in 3.2 to investigate the consequences of symmetry, which are particularly important in the vicinity of a symmetry changing instability, and can, in fact, be used to determine the general form of the stationary distribution. The procedure is completely analogous to the Landau theory of second order phase transitions [17].

Having determined the stationary distribution, it is still not possible to reduce the dynamic theory of stationary nonequilibrium states to the equilibrium theory. In equilibrium theory there exists a general, unique connection between the stationary distribution and the dynamics of the system, since both are determined by the same Hamiltonian. This connection is lacking in the nonequilibrium theory. As is shown in 2.2 the probability current in the stationary state has to be known in addition to the stationary distribution, in order to determine the dynamics. This difference from equilibrium theory is corroborated in 3.3 by looking at the generalization of the fluctuation dissipation theorem for stationary nonequilibrium states. As in equilibrium theory it is possible to express the linear response of the system in terms of a two-time correlation function. It is not possible, however, to calculate this correlation function and the stationary distribution from one Hamiltonian.

In Section 4 systems with the property of detailed balance are considered. In 4.2 and 4.3 it is shown, that, for such systems, there exists an analogy to thermal equilibrium states, with respect to their dynamic behaviour. For such systems, a phenomenological approach can be used to determine the dynamics from the stationary distribution. In 4.1 and 4.2 the conditions for the validity of detailed balance are examined. In particular, it is found, that a detailed balance condition holds in the vicinity of symmetry changing instabilities, when only a single mode is unstable. If several modes become unstable simultaneously, the presence of detailed balance depends on the existence of symmetries between these modes.

In part B the general phenomenological theory is applied to various optical examples. Some common characteristics of these examples and an outline of the alternative microscopic theory of the optical instabilities is set forth in Section 5.

Section 6 is devoted to various examples from laser theory. The laser presents an example of a system, which shows various instabilities in succession, each of which is connected with a new change in symmetry. In the Sections 6.1, 6.2, 6.3 we consider these transitions by means of the phenomenological theory. In Section 6.4 we consider as an example for a spatially extended system light propagation in a one dimensional laser medium.

The fluctuations near the instability leading to single mode laser action have been investigated experimentally in great detail [10, 18]. The experimental results were found to be in complete agreement with the results obtained by a Fokker-Planck equation, which was derived from a microscopic, quantized theory [8, 19]. In Section 6.1 we obtain from our phenomenological approach the same Fokker-Planck equation, and hence, all the experimentally confirmed results of the microscopic theory. The number of parameters which have to be determined by fitting the experimental results is the same, both, in the microscopic theory and in the phenomenological theory.

In Section 7 the phenomenological theory is applied to the most important class of instabilities in nonlinear optics, i.e. those which are connected with second order parametric scattering. The special case of subharmonic generation (cf. 7.2) presents an example where the symmetry, which is changed at the instability, is discontinuous, as in the example in Fig. 2. In this case fluctuations lead to small oscillations around the stable state and to discrete jumps between the degenerate stable states. The continuous phase diffusion occurs only in the non-degenerate parametric oscillator, treated in 7.1.

In Section 8 higher order scattering processes and multimode effects are considered by combining the microscopic and the macroscopic approach. The microscopic theory is used to derive the drift and diffusion terms of the Fokker-Planck equation in 8.2. The macroscopic theory is used to identify the conditions for the validity of detailed balance in 8.1 and to calculate the stationary distribution in 8.3, making use of the results of Section 4. The result, obtained in this way, is very general and makes it possible to discuss many special cases, some of which are considered in 8.4.

Throughout part B we try to make contact with the microscopic theory of the various instabilities. This comparison gives in some cases an independent check of the results of the phenomenological theory. On the other hand, this comparison is also useful for a further understanding of the microscopic theory, since it shows clearly which phenomenona have a microscopic origin and which not. We expect, therefore, that a combination of both, the phenomenological and the microscopic theory, will prove to be most useful in the future.

2. Continuous Markoff Systems

A general framework for the description of open systems is obtained by making some general assumptions. In this paper, we are only interested in macroscopic systems, which can be described by a small number of macroscopic variables, changing slowly and continuously in time. Therefore, the natural frame for a dynamic description is furnished by a Fokker-Planck equation, which combines drift and diffusion in a natural way. For reviews of the properties of this equation see, e.g., [20, 21]. Various equivalent formulations of the equations of motion are given in Sections 2.1 – 2.3. They allow us to consider a stationary nonequilibrium system as a generalization of an equilibrium system from various points of view. This comparison with equilibrium theory is useful and necessary in order to construct a phenomenological theory.

2.1. Basic Assumptions and Equations of Motion

Let us consider a system whose macroscopic state is completely described by a set of n variables

$$\{w\} = \{w_1, w_2, \ldots, w_i, \ldots, w_n\} . \tag{2.1}$$

Examples of such variables are: a set of mode amplitudes in optics, a set of concentrations in chemistry or a complete set of variables describing the hydrodynamics of some given system. On a macroscopic level of description neglecting fluctuations, the variables $\{w\}$ describe the state of the system.

A more detailed description takes into account, that the variables $\{w\}$ are, in general, fluctuating time dependent quantities. Thus, $\{w(t)\}$ forms an n-dimensional random process. The physical origin of the fluctuations can be quite different for various systems. Fluctuations may be imposed on the system from the outside by random boundary conditions or they may reflect a lack of knowledge about the exact state of the system, either because of quantum uncertainties (quantum noise) or because of the impossibility of handling a huge number of microscopic variables.

The random process formed by $\{w(t)\}$ may be characterized in the usual way by a set of probability densities

$$\begin{aligned} &W_1(\{w\}, t) \\ &W_2(\{w^{(2)}\}, t_2; \quad \{w^{(1)}\}, t_1) \\ &\cdots\cdots\cdots\cdots\cdots\cdots\cdots\cdots \\ &W_\nu(\{w^{(\nu)}\}, t_\nu; \quad \{w^{(\nu-1)}\}, t_{\nu-1}; \quad \cdots \{w^{(1)}\}, t_1) \end{aligned} \tag{2.2}$$

This hierarchy of distributions, instead of the set of variables (2.1), describes a state of the system, if fluctuations are important. W_ν is the ν-fold probability density for finding $\{w(t)\}$: near $\{w^{(1)}\}$ at the time $t=t_1$, near $\{w^{(2)}\}$ for $t=t_2, \ldots$, near $\{w^{(\nu)}\}$ for $t=t_\nu$.

As a first fundamental assumption we introduce the Markoff property of the random process $\{w(t)\}$, which is defined by the condition

$$\frac{W_\nu(\{w^{(\nu)}\}, t_\nu; \{w^{(\nu-1)}\}, t_{\nu-1}; \ldots \{w^{(1)}\}, t_1)}{W_{\nu-1}(\{w^{(\nu-1)}\}, t_{\nu-1}; \ldots \{w^{(1)}\}, t_1)} = P(\{w^{(\nu)}\} | \{w^{(\nu-1)}\}; t_\nu, t_{\nu-1}) \tag{2.3}$$

In (2.3) the conditional probability density P has been introduced, which only depends on the variables $\{w^{(\nu)}\}$, $\{w^{(\nu-1)}\}$ and the two times $t_\nu, t_{\nu-1}$.

From the Markoff assumption (2.3) it follows immediately that the whole hierarchy of distributions (2.2) is given, if W_1 and P are known. The condition (2.3) furthermore implies, that a Markoff process does not describe any memory of the system of states at times $t<t_0$ if at some time $t=t_0$ the system's state is specified by giving $\{w(t_0)\}$.

The physical content of the Markoff assumption is well known and may be summarized in the following way: It must be possible to separate the numerous variables, which give an exact microscopic description of the system, into two classes, according to their relaxation times. The first class, which is the set $\{w\}$, must have much longer relaxation times than all the remaining variables, which form the second class. The time scale of description is then chosen to be intermediate to the long and the short relaxation times. Then, clearly, all memory effects are accounted for by the variables $\{w\}$ and it is adequate to assume that they form a Markoff process.

As a consequence of Eq. (2.3) the probability density W_1 obeys the equation

$$W_1(\{w^{(2)}\}, t_2) = \int \{d\,w^{(1)}\}\, P(\{w^{(2)}\}|\{w^{(1)}\}; t_2, t_1)\, W_1(\{w^{(1)}\}, t_1) \tag{2.4}$$

which is obtained by integrating the expression for W_2, following from Eq. (2.3), over $\{w^{(1)}\}$.

A second fundamental assumption is the stationarity of the random process $\{w(t)\}$. This assumption implies, that all external influences on the system are time independent on the adopted time scale of description. It implies, furthermore, that the classification of the system's variables as slowly and rapidly varying quantities must be preserved during the evolution of the system. Owing to the assumption of stationarity the conditional distribution P in Eqs. (2.3), (2.4) depends only on the difference of the two times of its argument.

a) Fokker-Planck Equation

We simplify Eq. (2.4) by using the stationarity assumption. Furthermore, we write the integral Eq. (2.4) as a differential equation by taking $\tau = t_2 - t_1$ to be small, expanding P in terms of the averaged powers of $\{w^{(2)} - w^{(1)}\}$, and performing partial integrations. Eq. (2.4) then takes the form [1]

$$\dot{W}_1(\{w\}, t) = \sum_{s=1}^{\infty} \frac{(-1)^s}{s!} \left[\frac{\partial^s K_{i_1, i_2, \ldots, i_s}(\{w\})\, W_1(\{w\}, t)}{\partial w_{i_1}\, \partial w_{i_2} \ldots \partial w_{i_s}}\right] \tag{2.5}$$

where the coefficients $K\ldots$ are given by

$$\begin{aligned} K_{i_1, i_2, \ldots, i_s}(\{w\}) = \lim_{\tau \to 0}\, (1/\tau)\, \langle (w_{i_1}(t+\tau) - w_{i_1}(t))\, (w_{i_2}(t+\tau) - w_{i_2}(t)) \\ \ldots (w_{i_s}(t+\tau) - w_{i_s}(t)) \rangle \{w(t)\} = \{w\}\,. \end{aligned} \tag{2.6}$$

The angular brackets define the mean values of the enclosed quantities. The coefficients $K\ldots$ do not depend on t, due to the stationarity assumption [2]. The function $P(\{w^{(2)}\}|\{w^{(1)}\}; \tau)$, whose expansion in terms of the moments (2.6) led to Eq. (2.5), is recovered from Eq. (2.5) as its Green's function solution obeying the initial condition

$$P(\{w^{(2)}\}|\{w^{(1)}\}; 0) = \prod_{(i)} \delta(w_i^{(2)} - w_i^{(1)}) \tag{2.7}$$

Equations of the structure (2.5) are well known in many different fields of physics, where they were derived from microscopic descriptions.

[1] Summation over repeated indices is always implied, if not noted otherwise.

[2] Note, that Eq. (2.5) with time dependent $K\ldots$ holds even for non-Markoffian processes [20].

Most recently, perhaps, Eq. (2.5) has been derived in quantum optics for electromagnetic fields interacting with matter (cf. [8]).

Owing to the appearance of derivatives of arbitrarily high order, Eq. (2.5) is in most cases too complicated to be solved in this form. In the following, we simplify Eq. (2.5) by dropping all derivatives of higher than the second order. Eq. (2.5) then acquires the basic structure of a Fokker-Planck equation. Mathematically speaking, the Markoff process Eq. (2.5) is reduced to a continuous Markoff process in this way.

A physical basis for the truncation of Eq. (2.5) after the second order derivatives can often be found by looking at the dependence of the coefficients $K\ldots$ on the size of the system. To this end the variables $\{w\}$ have to be rescaled in order to be independent of the system's size. If the fluctuations described by the coefficients $K\ldots$ have their origin in microscopic, non-collective events, the coefficients of derivatives of subsequent orders in Eq. (2.5) decrease in order of magnitude by a factor increasing with the size of the system.

As a zero order approximation we obtain from Eq. (2.5)

$$\partial W_1/\partial t = -\partial K_i(\{w\})\, W_1/\partial w_i\,. \tag{2.8}$$

This equation can easily be solved, if the solutions of its characteristic equations

$$\dot{w}_i = K_i(\{w\}) \tag{2.9}$$

are known. Eq. (2.8) describes a drift of W_1 in the $\{w\}$-space along the characteristic lines given by Eq. (2.9). In this drift approximation fluctuations are introduced only by the randomness, which is contained in the initial distribution. In order to describe a fluctuating motion of the system, we have to include the second order derivative terms in Eq. (2.5); this leads to the Fokker-Planck equation

$$\partial W_1/\partial t = -\partial K_i(\{w\})\, W_1/\partial w_i + \tfrac{1}{2}\partial^2 K_{ik}(\{w\})\, W_1/\partial w_i \partial w_k\,. \tag{2.10}$$

The second order derivatives describe a generalized diffusion in $\{w\}$-space. The diffusion approximation (2.10) of Eq. (2.5) is adopted in all the following.

From Eq. (2.6) the diffusion matrix $K_{ik}(\{w\})$ is obtained symmetric and non-negative. We also assume in the following that the inverse of K_{ik} exists. Singular diffusion matrices can be treated as a limiting case.

Eq. (2.10) has to be supplemented by a set of initial boundary conditions. The initial condition is given by the distribution W_1 for a given time. The special choice (2.7) gives P as a solution of Eq. (2.10). As boundary conditions we may specify W_1 and its first order derivatives at the boundaries. We will assume "natural boundary conditions" in the following, i.e., the vanishing of W_1 and its derivatives at the boundaries.

The conditional distribution P also satisfies, besides Eq. (2.10), the adjoint equation, which is called the backward equation. It is obtained by differentiating the relation

$$W_1(\{w\},t)=\int\{\mathrm{d}w'\}\,P(\{w\}|\{w'\};\tau)\,W_1(\{w'\},t-\tau) \tag{2.11}$$

with respect to τ and using Eq. (2.10) to express the time derivative of W_1 on the right hand side of this equation. The differential operations on $W_1(\{w'\},t-\tau)$ are then transferred to P by partial integrations, using the natural boundary conditions. Finally, since W_1 is an arbitrary distribution, integrands can be compared to yield

$$[\partial/\partial\tau-K_i(\{w'\})\,\partial/\partial w_i'-\tfrac{1}{2}K_{ik}(\{w'\})\,\partial^2/\partial w_i'\partial w_k']\,P(\{w\}|\{w'\};\tau)=0\,. \tag{2.12}$$

This equation will be used in Section 4.2.

b) Langevin Equations

Instead of Eq. (2.10) one may use a set of equations of motion for the time dependent random variables $\{w(t)\}$ themselves. These are the Langevin equations, which are stochastically equivalent to the equation for the probability distributions W_1 or P, in the sense that the final results for all averaged quantities are the same. The Langevin equations corresponding to the Fokker-Planck equation (2.10) take the form[3] [20]:

$$\dot{w}_i=K_i(\{w\})+F_i(\{w\},t) \tag{2.13}$$

with

$$F_i(\{w\},t)=-\tfrac{1}{2}(\partial g_{ij}(\{w\})/\partial w_k)\,g_{kj}(\{w\})+g_{ik}(\{w\})\,\xi_k(t)\,. \tag{2.14}$$

The $(n\times n)$-matrix $g_{ik}(\{w\})$ has to obey the $n(n+1)$ relations

$$g_{ik}g_{jk}=K_{ij} \tag{2.15}$$

and is arbitrary otherwise.

The quantities $\xi_k(t)$ are Gaussian, δ-correlated fluctuating quantities with the averages

$$\langle\xi_i(t)\rangle=0\,, \tag{2.16}$$

$$\langle\xi_i(t)\,\xi_j(t+\tau)\rangle=\delta_{ij}\,\delta(\tau)\,. \tag{2.17}$$

The higher order correlation functions and moments of the $\{\xi\}$ are determined by (2.16), (2.17) according to their Gaussian properties.

[3] For K_{ij} independent of $\{w\}$ the Langevin equations are equivalent to the Fokker-Planck equation. Otherwise the correspondence is approximate only (cf. [20]).

A characteristic feature of all Langevin equations, which also occurs in Eq. (2.13), is the separation of the time variation into a slowly varying and a rapidly varying part. In the present case this separation is not unique, since we may impose another $n(n-1)/2$ independent conditions on g_{ij}, besides the $n(n+1)/2$ relations (2.15), in order to fix its n^2 elements completely. Usually, these relations are chosen to make g_{ij} symmetric

$$g_{ij} = g_{ji} \tag{2.18}$$

which implies, that now the i'th noise source is coupled to w_j in the same way as the j'th noise source is coupled to w_i. This condition is by no means compelling and can be replaced by other conditions, if this happens to be convenient[4]. While this would change g_{ij} and the mean value of the fluctuating force

$$\langle g_{ij}(\{w\})\, \xi_j(t)\rangle = \tfrac{1}{2}(\partial g_{ij}(\{w\})/\partial w_k)\, g_{kj}(\{w\}) \tag{2.19}$$

it would leave unchanged all results for $\{w(t)\}$, after the average has been performed. This may be simply proven by deriving Eq. (2.10) from Eq. (2.13) [20].

Physically, the appearance of a coupling of the $\{w(t)\}$ to a set of Gaussian random variables with very short correlation times reflects the coupling of the macroscopic variables to a large number of statistically independent, rapidly varying microscopic variables. Therefore, Eq. (2.13) gives a very transparent mathematical expression to our basic physical assumptions.

2.2. Nonequilibrium Theory as a Generalization of Equilibrium Theory[5]

The equations of motion obtained in the last section can be compared with familiar equations of equilibrium theory. The Fokker-Planck equation (2.10) may be written as a continuity equation for the probability density W_1 in the general form

$$\partial W_1(\{w\}, t)/\partial t + \partial(r_i(\{w\}, t)\, W_1(\{w\}, t))/\partial w_i = 0\,. \tag{2.20}$$

In Eq. (2.20) we introduced the drift velocity $\{r(\{w\}, t)\}$ in $\{w\}$-space. In order to establish a connection with equilibrium theory we define a "potential" $\phi(\{w\}, t)$ by putting

$$W_1(\{w\}, t) = N \exp(-\phi(\{w\}, t))\,. \tag{2.21}$$

[4] For $n>2$ a possible condition is $\partial g_{ij}/\partial w_i = 0$ for all j, in which case some of the following expressions are simplified considerably.

[5] By equilibrium theory we mean the theory of thermal equilibrium and the linearized theories in the vicinity of thermal equilibrium.

Here, N is a normalization constant, which is independent of $\{w\}$ and t. Comparing now Eq. (2.20) with Eq. (2.10) and using Eq. (2.21) we may express the drift coefficient $K_i(\{w\})$ in terms of the newly defined quantities ϕ and $\{r\}$. We obtain

$$K_i - \tfrac{1}{2}\partial K_{ik}/\partial w_k = -\tfrac{1}{2}K_{ik}\,\partial\phi/\partial w_k + r_i\,. \tag{2.22}$$

The left hand side of Eq. (2.22) represents the total drift, as can be seen by writing Eq. (2.10) in the form

$$\partial W_1/\partial t + \frac{\partial}{\partial w_i}\left(K_i - \frac{1}{2}\partial K_{ik}/\partial w_k - \frac{1}{2}K_{ik}\frac{\partial}{\partial w_k}\right)W_1 = 0\,. \tag{2.23}$$

Eq. (2.22) shows, that the total drift can generally be decomposed into two parts. The first part is connected with the first order derivatives of the potential $\phi(t)$. The second part is the drift velocity of the probability current which satisfies the continuity Eq. (2.20). The decomposition (2.22) holds for all potentials $\phi(t)$ and velocities $\{r(t)\}$ which together satisfy Eq. (2.20) at a given time. Of special interest is the pair $\phi^s(\{w\})$ and $\{r^s(\{w\})\}$ which solves Eq. (2.20) in the stationary state with $\partial W_1^s/\partial t = 0$. By introducing the decomposition (2.22) into the Langevin equations we obtain

$$\dot{w}_i = r_i^s - \tfrac{1}{2}K_{ik}\,\partial\phi^s/\partial w_k + \tfrac{1}{2}g_{ik}\,(\partial g_{jk}/\partial w_j + \xi_k(t))\,. \tag{2.24}$$

The decomposition (2.22) is well known from the theory of systems near thermal equilibrium, where it acquires a special meaning. There, the decomposition (2.22) simultaneously is a decomposition of the total drift into two parts which differ in their time reversal properties. The first part of the drift in Eq. (2.22) describes the irreversible processes. The expressions $-\tfrac{1}{2}K_{ik}\,\partial\phi^s/\partial w_k$ represent the familiar set of phenomenological relations giving the irreversible drift terms as linear functions of the thermodynamic forces, defined by the derivatives of a thermodynamic potential [13]. The coefficients K_{ik} are then the Onsager coefficients in these relations. The fact that they also give the second order correlation coefficients of the fluctuating forces is a familiar relation for thermal equilibrium. The remaining part of the drift is associated with reversible processes, described by some Hamiltonian. The continuity Eq. (2.20), satisfied by this part, is then simply an expression for the conservation of energy in the form of a Liouville equation.

Unfortunately, such a simple physical interpretation of the two different parts of the drift is not possible, in general, for nonequilibrium states. There, both parts contain contributions from reversible and irreversible processes. Eq. (2.22) is then no help for calculating the potential ϕ^s, and the stationary distribution W_1^s from the drift and diffusion coefficients.

In all cases, however, in which the potential ϕ^s, the velocity $\{r^s\}$ and the diffusion coefficients K_{ik} are known by other arguments (e.g. by symmetry). Eq. (2.22) is useful to determine the drift $K_i(\{w\})$. This gives the key for a phenomenological analysis of the dynamics of stationary nonequilibrium systems in cases in which symmetry arguments play an important role (cf. section 3).

2.3. Generalization of the Onsager-Machlup Theory

In this section we put the equations obtained in 2.1 on a common basis with the phenomenological theory of thermodynamic fluctuations. While this is useful from a systematic point of view, it is not necessary for an understanding of the other sections.

A set of Langevin equations of the form (2.13) has been used by Onsager and Machlup [14] as a starting point for a general theory of time dependent fluctuations of thermodynamic variables. However, an essential restriction of their theory was the assumption of the linearity of Eqs. (2.13). The same assumption has also been used by a number of subsequent authors [15, 16], although the necessity for a generalization of the Onsager Machlup theory to include nonlinear processes was emphasized [16].

In this section we shall give such a generalization, starting from Eqs. (2.13) and allowing for nonlinear functions $K_i(\{w\})$ and $g_{ij}(\{w\})$. This generalization will serve the two purposes: first, showing in which limit the usual thermodynamic fluctuation theory is contained in the present formulation and second, showing the limits of the Onsager Machlup formulation of fluctuation theory for general Langevin Eqs. (2.13). An essential point of the Onsager Machlup theory is to consider probability densities for an entire path $\{w(t)\}$ in some given time interval, rather than for $\{w(t_\nu)\}$ at a given time t_ν. The probability density for an entire path is obtained from the hierarchy (2.2) in the limit in which the differences between different times go to zero. In this limit we obtain a probability density functional $W_\infty[\{w\}]$ of the paths $\{w(t)\}$ which may be viewed as a function of the infinite number of variables $\{w(t)\}$ taken at all times in some given time interval $t_1 \leqq t \leqq t_2$. The Onsager Machlup theory can now be characterized by the postulates [16] that

i) $\{w(t)\}$ is a stationary Markoff process, and

ii) the probability density *functional* $W_\infty[\{w\}]$ is determined by a *function* $O(\{w(t)\}, \{\dot{w}(t)\})$ in the following way:

$$W_\infty[\{w\}] = G(F_{2,1}) \tag{2.25}$$

where F_{21} is defined by the integral

$$F_{21} = \int_{\{w^{(1)}(t_1)\},t_1}^{\{w^{(2)}(t_2)\},t_2} \mathrm{d}t\, O(\{w(t)\},\{\dot{w}(t)\}). \tag{2.26}$$

G in Eq. (2.25) is a nonnegative but otherwise arbitrary function. It can be determined by the following argument. From the first postulate we infer, that the conditional probability density P obeys the relation

$$\begin{aligned} &P(\{w^{(\nu)}\}|\{w^{(1)}\};t_\nu - t_1) \qquad (2.27)\\ &= \int \{\mathrm{d}w^{(\nu-1)}\}\, P(\{w^{(\nu)}\}|\{w^{(\nu-1)}\};t_\nu - t_{\nu-1})\, P(\{w^{(\nu-1)}\}|\{w^{(1)}\};t_{\nu-1}-t_1). \end{aligned}$$

On the other hand P is given in terms of W_∞ by the functional integral

$$P(\{w^{(\nu)}\}|\{w^{(1)}\};t_\nu - t_1) = \int [\{\mathrm{d}w\}]\, G\left(\int_{\{w^{(1)}(t_1)\},t_1}^{\{w^{(\nu)}(t_\nu)\},t_\nu} O(\{w(t)\},\{\dot{w}(t)\})\,\mathrm{d}t\right) \tag{2.28}$$

where the integration runs over all paths passing through the indicated boundary values. The integrand in Eq. (2.28) could also be expressed as $G(F_{\nu 1})$. Taking Eqs. (2.27) and (2.28) together, we obtain the relation

$$\int [\{\mathrm{d}w\}]\,(G(F_{\nu,1}) - G(F_{\nu,\nu-1})\,G(F_{\nu-1,1})) = 0\,. \tag{2.29}$$

Since this equation must be fulfilled for all choices of the intermediate boundary of integration $\{w^{(\nu-1)}(t_{\nu-1})\}$, Eq. (2.29) is a relation for the non-negative function G, which has the simple structure

$$g(F_1 + F_2) = g(F_1)\, g(F_2)\,. \tag{2.30}$$

The unique, nonsingular and nontrivial solution of Eq. (2.30) has the form

$$g(F) \sim \exp(aF)\,. \tag{2.31}$$

By measuring the function O in appropriate units, we may take $a = -1$ and obtain

$$W_\infty[\{w\}] \sim \exp[-\int \mathrm{d}t\, O(\{w(t)\},\{\dot{w}(t)\})] \tag{2.32}$$

which determines W_∞ up to a normalization constant, which will not depend on $\{w\}$, $\{\dot{w}\}$.

An expression of the form (2.32) is useful as a starting point of fluctuation theory, as was first noted by Onsager and Machlup. Eq. (2.32) establishes for time dependent fluctuations a relation between a probability density and an additive quantity, the Onsager Machlup function O. O has thermodynamic significance, since it can be related to the entropy production. Therefore, Eq. (2.32) is the time dependent analogue to the familiar relation between probability density and entropy which holds in the static case. In addition, Eq. (2.32) is valuable, because it contains in a concise form the most complete information on the paths $\{w(t)\}$. Hence, the Onsager Machlup function O plays a role in fluctuation theory, which is similar to the role of the Lagrangian in mechanics.

We determine now the Onsager Machlup function which is equivalent to the equations of motion (2.13). The Onsager Machlup function of $\{\xi(t)\}$, introduced in (2.14), may be written down immediately, by using Eqs. (2.16), (2.17). We obtain

$$W_\infty[\{\xi\}] = \lim_{\Delta t\to 0} \prod_{\nu=1}^{N} (\sqrt{\Delta t/2\pi} \cdot \mathrm{d}\xi(t_\nu)) \exp\left[-\tfrac{1}{2}\int_{t_0}^{t_1} \mathrm{d}t\, \xi_i(t)\, \xi_i(t)\right] \tag{2.33}$$

where $t_0 \leqq t \leqq t_1$ is some given time interval and

$$t_\nu = t_0 + \nu t_1/N; \qquad N = (t_1 - t_0)/\Delta t \tag{2.34}$$

is a discrete time scale which becomes continuous in the limit $\Delta t \to 0$, $N \to \infty$. From (2.33) we obtain

$$O(\{\xi(t)\}) = \tfrac{1}{2}\xi_i(t)\, \xi_i(t)\,. \tag{2.35}$$

From Eq. (2.33) we may derive an expression for $W_\infty[\{w\}]$, since Eq. (2.13) defines a mapping of both functionals on each other. The probability

$$W_\infty[\{\xi\}]\,[\{\mathrm{d}\xi\}] = W_\infty[\{w\}]\,[\{\mathrm{d}w\}] \tag{2.36}$$

has a physical meaning and is an invariant of this mapping. The volume elements in function space are connected by the Jacobian of the mapping (2.13)

$$[\{\mathrm{d}\xi\}] = D(\{w\})\,[\{\mathrm{d}w\}]\,. \tag{2.37}$$

Since the mapping (2.13) is nonlinear in our case, the Jacobian is not merely a constant, as in the Onsager Machlup theory, which could be absorbed into the normalization constant, but it rather is dependent on $\{w\}$ and has to be calculated. This can be done in a conventional way by introducing a discrete time scale, Eq. (2.34), and passing to the continuous limit at the end of the calculations. The discretization of Eq. (2.13) has to be done with some care, introducing only errors of the order $(\Delta t)^2$, in order to obtain the correct continuous limit $\Delta t \to 0$. We skip the lengthy but elementary calculation and give immediately the result for the Jacobian

$$\begin{aligned} D(\{w\}) = &\left\{\lim_{\Delta t\to 0} \prod_{(\nu)} [\Delta t \sqrt{\mathrm{Det}(K_{ik}^{(\nu)})}]^{-1}\right\} \\ &\exp\left[-\tfrac{1}{2}\int_{t_0}^{t_1} \cdot \mathrm{d}t(\partial K_j/\partial w_j - \tfrac{1}{2}\partial^2 K_{jk}/\partial w_j\,\partial w_k\right. \\ &\left. + (\partial g_{jk}/\partial w_j)\cdot g_{kn}^{-1}(\dot{w}_n - K_n + \tfrac{1}{2}\partial K_{nl}/\partial w_l) + \tfrac{1}{2} g_{jk}\,\partial^2 g_{lk}/\partial w_j\,\partial w_l)\right]. \end{aligned} \tag{2.38}$$

$K_{ik}^{(\nu)}$ is defined by

$$K_{ik}^{(\nu)} = K_{ik}(\{w(t_\nu)\}). \tag{2.39}$$

We can now write down the complete functional $W_\infty[\{w\}]$, by introducing the mapping (2.13) into Eq. (2.33) and taking into account Eqs. (2.36) – (2.38).

$$W_\infty[\{w\}][\{dw\}] = \left\{ \lim_{\Delta t \to 0} \prod_{(\nu)} \{dw(t_\nu)\} \left[2\pi \Delta t \cdot \mathrm{Det}(K_{ik}^{(\nu)})\right]^{-1/2} \right\} \cdot \exp\left(- \int_{t_0}^{t_1} dt\, O(\{w(t)\}, \{\dot{w}(t)\}) \right) \tag{2.40}$$

The Onsager Machlup function is obtained as

$$O(\{w(t)\}, \{\dot{w}(t)\}) = \tfrac{1}{2}(\dot{w}_i - K_i + \tfrac{1}{2}\partial K_{ik}/\partial w_k)\, K_{ij}^{-1}(\dot{w}_j - K_j + \tfrac{1}{2}\partial K_{jl}/\partial w_l) + \tfrac{1}{2}\partial(K_i - \tfrac{1}{2}\partial K_{ij}/\partial w_j)/\partial w_i + \tfrac{1}{8}(\partial^2 K_{ij}/\partial w_i\, \partial w_j - (\partial g_{ij}/\partial w_k)(\partial g_{kj}/\partial w_i)). \tag{2.41}$$

Eqs. (2.40), (2.41) generalize the result for linear processes in two ways. First, Eq. (2.41) contains a correction term which comes from the nonlinearity of the total drift $K_i(\{w\}) - \frac{1}{2}\partial K_{ik}(\{w\})/\partial w_k$. Secondly, the dependence of the diffusion coefficients $K_{ik}(\{w\})$ on the variables alters the form of the functional (2.40). Eq. (2.40) shows, in fact, that the second postulate of the Onsager-Machlup theory is no longer valid if the diffusion coefficients are functions of the variables $\{w\}$, since the Onsager Machlup function alone does no longer determine the probability density functional.

The expressions (2.40), (2.41) can be used as a starting point to derive in a systematic way the equations of the preceding sections. We indicate very briefly how this can be done. The conditional probability density $P(\{w^{(1)}\}|\{w^{(0)}\}, t_2 - t_1)$ is given in terms of O by the functional integral

$$P(\{w^{(1)}\}|\{w^{(0)}\}, t_1 - t_0) = \int_{\{w^{(0)}(t_0)\}}^{\{w^{(1)}(t_1)\}} W_\infty[\{w\}][\{dw\}] \tag{2.42a}$$

with Eq. (2.40). This functional integral has a pronounced analogy to the path integrals introduced by Feynman into quantum mechanics [22]. In fact, it was shown by Feynman that the Green's function G of the Schrödinger equation for a particle of mass m moving from a point in space $\{x^{(0)}\}$ at time t_0 to a point $\{x^{(1)}\}$ at time t_1, can be obtained as the functional integral

$$G(\{x^{(1)}\}|\{x^{(0)}\}, t_1 - t_0) = \int_{\{x^{(0)}(t_0)\}}^{\{x^{(1)}(t_1)\}} \lim_{\Delta t \to 0} \prod_{(\nu)} \mathrm{d}x^{(\nu)} \cdot (2\pi \Delta t m^{-1} \hbar i)^{-1/2} \cdot \exp\left[i\hbar^{-1} \int_{t_0}^{t_1} \mathrm{d}t \cdot L(\{x\}, \{\dot{x}\}) \right] \quad (2.42b)$$

where L is the Lagrangian of the particle. From this formal analogy a number of interesting results immediately follow. O is, in fact, the analogue of a Lagrangian for the motion in $\{w\}$-space. Once O is known, the Fokker-Planck equation can be derived in analogy to the derivation of the Schrödinger equation in the Feynman theory. This analogy of the Fokker-Planck equation and the Schrödinger equation proved to be very useful in laser theory [19] and many different fields of statistical mechanics (cf. the papers by Montroll, Kawasaki, Zwanzig in [23]). The analogue of the classical limit of a very heavy particle ($m \to \infty$) in quantum mechanics is, in our case, the limit of vanishing fluctuations $K_{ik} \to 0$. In this limit the "Lagrangian" equations

$$\frac{\mathrm{d}}{\mathrm{d}t} \frac{\partial O}{\partial \dot{w}_k} - \frac{\partial O}{\partial w_k} = 0 \quad (2.43)$$

give an adequate description. For nonvanishing fluctuations, but constant diffusion coefficients K_{ik}, these equations still remain valid if they are averaged over the fluctuations, in analogy to Ehrenfest's theorem of quantum mechanics.

3. The Stationary Distribution

In this section we will consider some general properties of the stationary state in descriptions which either neglect or include fluctuations. Of particular interest are the symmetry changing transitions between different branches of states, which are caused by instabilities of the system. In the first subsection we give a discussion of various stability concepts and obtain several results on the stability of the stationary state. In the second subsection we consider some consequences of symmetry for the stationary distribution. The results of these subsections are quite analogous to results of equilibrium theory. It will become clear that a close analogy exists between second order phase transitions and symmetry changing transitions between different branches of stationary nonequilibrium states, and that a phenomenological approach can be used to obtain the stationary distribution in the vicinity of the instability.

The limits of the analogy are shown in the third subsection, where we discuss the dissipation fluctuation theorem for stationary nonequilibrium states.

3.1. Stability and Uniqueness

In Section 2 we introduced two different descriptions for the "state of the system". The first was given by a set of numbers $\{w\}$, Eq. (2.1), the second was given by a set of probability densities, Eq. (2.2). With both descriptions we may associate a definition of the stationary state and of the stability of the stationary state.

a) Stability of a Single State

Let us first deal with the description furnished by the set of numbers (2.1). This description is adequate if fluctuations can be neglected. A stationary state is obtained if

$$\{w^s(t)\} = \{w^s(t+T)\} \tag{3.1}$$

is either constant or periodic in time with some constant period $T \geqq 0$. The probability distribution, corresponding to (3.1) is

$$W_1^s = \prod_{(i)} \delta(w_i - w_i^s(t)) \tag{3.2}$$

which changes periodically in time. The stationary distribution, which one obtains as a limit for very small K_{ik}, is not Eq. (3.2) but rather the time average

$$W_1^s = T^{-1} \int_t^{t+T} \mathrm{d}t \prod_{(i)} \delta(w_i - w_i^s(t)) \tag{3.3}$$

which defines a time independent surface in $\{w\}$-space, rather than a moving point, like (3.2)[6]. The dynamics is described, in the present case, by the drift approximation Eq. (2.8) of the Fokker-Planck equation, or by the Langevin equations in the same limit, which, according to Eq. (2.24), may be put into the form

$$\dot{w}_i = -\tfrac{1}{2} K_{ij} \partial \phi^s / \partial w_j + r_i^s \,. \tag{3.4}$$

The potential ϕ^s is given by the stationary distribution

$$W_1^s \sim \exp(-\phi^s) \,. \tag{3.5}$$

[6] If several stable states (3.1) coexist, the limiting distribution (3.3) is distributed over several surfaces.

The stationary drift velocity $\{r^s\}$ satisfies the equation (cf. Eqs. (2.20) (2.22))

$$\partial r_i^s W_1^s / \partial w_i = 0 . \tag{3.6}$$

Since $\{r^s\}$ is the stationary drift velocity, $\{w^s\}$ has to fulfill the equation

$$\dot{w}_i^s = r_i^s(\{w^s\}) . \tag{3.7}$$

By comparison with Eq. (3.4) we find

$$\partial \phi^s(\{w^s\}) / \partial w_j^s = 0 \tag{3.8}$$

which is satisfied for all states of maximum or minimum probability.

In order to analyze the stability of these states we distinguish two cases. In the first case

$$r_i^s(\{w\}) \, \partial \phi^s(\{w\}) / \partial w_i = 0 . \tag{3.9}$$

In the second case Eq. (3.9) does not hold. In the latter case $\{r^s\}$ has a component orthogonal to surfaces of equal potential ϕ^s, and no general prediction about the stability of the stationary state can be made.

If (3.9) is satisfied, ϕ^s can be used as a Lyapunoff function [24] for Eq. (3.4), since the total time derivative of ϕ^s is given by

$$\mathrm{d}\phi^s/\mathrm{d}t = (\partial \phi^s / \partial w_i) \, \dot{w}_i = -\tfrac{1}{2} K_{ij} (\partial \phi^s / \partial w_i)(\partial \phi^s / \partial w_j) \leqq 0 \tag{3.10}$$

and is always negative, except when condition (3.8) is fulfilled, when it is zero. Here we made use of the positive definiteness of the diffusion matrix. In a neighbourhood of stationary trajectories connecting points of maximum probability density (minimum ϕ^s) we have

$$\phi^s - \phi^s_{\min} \geqq 0 \tag{3.11}$$

$$\mathrm{d}(\phi^s - \phi^s_{\min})/\mathrm{d}t \leqq 0 . \tag{3.12}$$

If $\{w^s\}$ is a local, non-degenerate minimum of ϕ^s, the $>$ sign in (3.11) holds for $\{w\} \neq \{w^s\}$. In this case $\phi^s - \phi^s_{\min}$ has all the required properties of a Lyapumoff function and the state $\{w^s\}$ is found to be stable. For $\{w^s\}$ independent of time, it follows from Eq. (3.7) that $\{r^s(\{w^s\})\} = 0$.

In the case where the minimum of ϕ^s are continuously degenerate, there are states in the neighbourhood of each $\{w^s\}$ for which the equality sign in Eq. (3.11) holds. This is always realized, if $\{r^s(\{w^s\})\}$ is different from zero. Then the trajectory is still stable with respect to fluctuations towards states with lower W_1^s and higher ϕ_1^s. It is metastable with respect to fluctuations towards states with equal ϕ^s, which are either on different trajectories or on the same trajectory. Metastability of the latter kind leads to a diffusion of the phase of the periodic trajectories (3.1). The presence of fluctuations, even if they are very small, thus completely changes the stability results for the stationary state. This will be considered further in the next subsection. Here we see that stable stationary states are associated with minima of ϕ^s and that several stable states may coexist simultaneously. The symmetry of the stationary states is given by the symmetry of the minima of ϕ^s.

b) Stability and Uniqueness of an Ensemble of States

If a statistical description of the system is used, a stationary state has to be defined by the condition, that the probability densities (2.2) depend on time in a periodic way. In particular

$$W_1^s(\{w\}, t+T) = W_1^s(\{w\}, t) \tag{3.13}$$

has to be constant or periodic in time with $T \geqq 0$. We will find that $T=0$ is the only possibility. The stability of the stationary state is now determined by the stability of the solutions (3.13) of Eq. (2.10). As was indicated in subsection *a*, even the slightest fluctuations change the stability considerations completely. The system of Eqs. (3.4) could have a manifold of stable solutions. If fluctuations are present, which allow the system to assume all values $\{w\}$, we find that generally only one stable probability density (3.13) describes the stationary state of the system. Hence, all the instabilities, which were possible in Eq. (3.4), are now buried, even in the slightest fluctuations. The instabilities manifest themselves only in the detailed form of the probability density W_1^s, as will be discussed in 3.2.

The proof of the stability and uniqueness of the stationary distribution of Eq. (2.10) has already been given by Lebowitz and Bergmann [25] under rather general conditions. We give here a short account of their proof. It consists in showing that the function

$$K(t) = \int \{dw\}\, W_1(\{w\}, t) \ln[W_1(\{w\}, t)/W_1^s(\{w\}, t)] \tag{3.14}$$

with the property

$$K(t) \begin{matrix} >0 & \text{for} & W_1 \not\equiv W_1^s \\ =0 & & W_1 \equiv W_1^s \end{matrix} \tag{3.15}$$

can only decrease in the course of time. The same function was employed in [5] for a general analysis of stationary nonequilibrium states. The property (3.15) can be shown by replacing $\ln(W_1/W_1^s)$ in Eq. (3.14) by $\ln(W_1/W_1^s) - 1 + W_1^s/W_1$, (which is possible because of the normalization condition for the probability densities) and using the inequality

$$\ln x^{-1} - 1 + x \begin{matrix} >0 & \text{for} & x>0, \quad x \neq 1, \\ =0 & \text{for} & x=1. \end{matrix} \tag{3.16}$$

The time variation of $K(t)$ is given by

$$K(t+\tau)-K(t) \tag{3.17}$$
$$=\int\{\mathrm{d}w\}\left[W_1(\{w\},t+\tau)\ln\frac{W_1(\{w\},t+\tau)}{W_1^s(\{w\},t+\tau)}-W_1(\{w\},t)\ln\frac{W_1(\{w\},t)}{W_1^s(\{w\},t)}\right]$$

which, by using Eq. (2.4), we may write as the double integral

$$K(t+\tau)-K(t)=\int\{\mathrm{d}w^{(1)}\}\,\{\mathrm{d}w^{(2)}\}\,P\{w^{(2)}\}|\{w^{(1)}\};t+\tau,t)\,W_1(\{w^{(2)}\},t)$$
$$\cdot[\ln Q+1-Q]\leqq 0 \tag{3.18}$$

with

$$Q=\frac{W_1(\{w^{(1)}\},t+\tau)\,W_1^s(\{w^{(2)}\},t)}{W_1^s(\{w^{(1)}\},t+\tau)\,W_1(\{w^{(2)}\},t)}\,. \tag{3.19}$$

If we assume that all points in $\{w\}$-space are connected with each other by some sequence of transitions, the equality sign in Eq. (3.18) holds if and only if $Q=1$, i.e.

$$\frac{W_1(\{w^{(1)}\},t+\tau)}{W_1^s(\{w^{(1)}\},t+\tau)}\equiv\frac{W_1(\{w^{(2)}\},t)}{W_1^s(\{w^{(2)}\},t)}=\text{const}\,. \tag{3.20}$$

The constant in Eq. (3.20) is 1 by normalization. This proves that $K(t)$ has the properties of a Lyapunoff functional for Eq. (2.10). It shows that all probability densities W_1^s approach each other in the course of time. If the limit exists, it is given by the stationary distribution W_1^s, which is unique and stable.

As a consequence, the periodic time behaviour, postulated for the stationary distribution W_1^s in (3.13), has to be specialized to time independence. Otherwise it would be possible to construct many different stationary solutions simply by shifting the time t by an arbitrary interval. More generally, it follows from the uniqueness of the stationary distribution, that W_1^s and ϕ^s have to be invariants of all symmetries of the system. Otherwise, many different stationary distributions could be generated by applying one of the symmetry transformations of the system. These transformations leave Eq. (2.10) unaltered, but would change the stationary distribution if it were not an invariant.

3.2. Consequences of Symmetry

The fact that the stationary distribution W_1^s is an invariant of all symmetry operations of the system has some interesting consequences, which are discussed now. For the case of weak fluctuations the distribution W_1^s will have rather sharp maxima. The behaviour of the system will then depend on the location of these maxima and on the behaviour of W_1^s in their vicinity. Both properties of W_1^s are determined by the symmetries of the system in the following way. Extrema of W_1^s appear on all points in $\{w\}$-space which are left invariant by some symmetry operation of the system (cf. Figs. 4, 5 point 0). Since W_1^s as a whole is an invariant, the vicinity of each extremum has to remain unchanged by the same symmetry operation which gave rise to the extremum. Therefore, a point in $\{w\}$-space which is invariant against all symmetry operations, has to be a local extremum of W_1^s with completely symmetric neighbourhood (Figs. 4, 5 point 0). Extrema with lower symmetry have a neighbourhood with lower symmetry. Such extrema must occur in de-

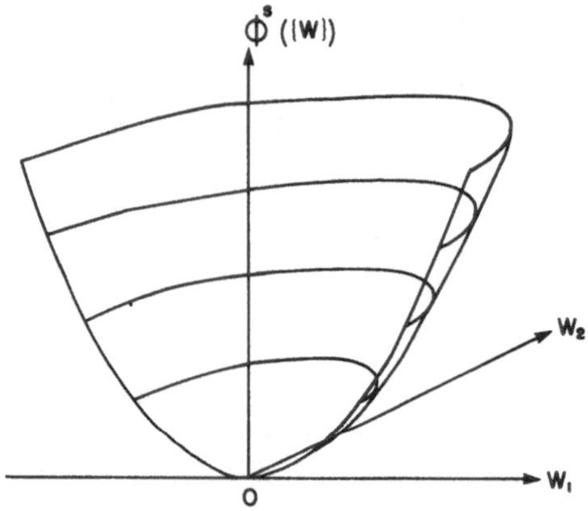

Fig. 4. The potential ϕ^s in the vicinity of a stable symmetric state 0 in a system with two-dimensional rotation symmetry

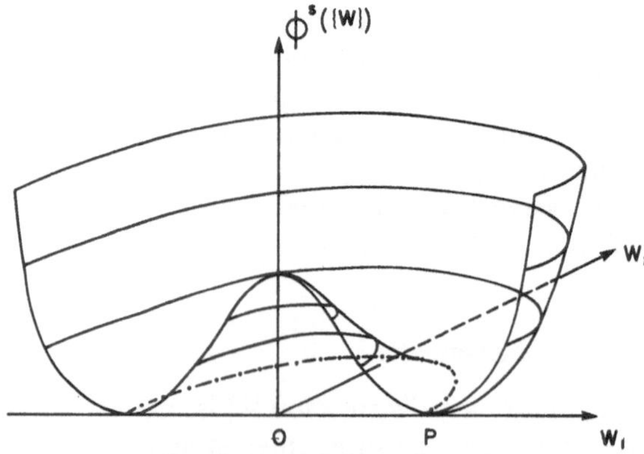

Fig. 5. The potential ϕ^s in the vicinity of a metastable state P with lower symmetry, for a system with two-dimensional rotation symmetry

generate groups. The degeneracy is either continuous on a whole surface in $\{w\}$-space (cf. Fig. 5 point P), or discontinuous (cf. Fig. 2, point P), depending on whether the symmetry broken by the extremum is continuous or discontinuous.

We consider now the reaction of the system, when we change the external forces acting on it. The external forces are described by a set of time independent parameters $\{\lambda\}$. It is always assumed that a change of $\{\lambda\}$ does not change the symmetries. Therefore, only the detailed forms of W_1^s and ϕ^s can depend on $\{\lambda\}$, but not their global symmetry (cf. Figs. 4, 5). In particular the location of the nondegenerate symmetric extrema of W_1^s cannot change. However, these fixed extrema can be transformed from minima into maxima and vice versa. These transformations are the cause for symmetry changing transitions. Consider, e.g., a highly symmetric maximum of W_1^s (point 0 in Fig. 4). As long as it retains its maximum property, a variation of $\{\lambda\}$ has only a small (quantitative) effect on the stationary state (3.1). Assume now that for some critical value $\{\lambda\}=\{\lambda_c\}$, the maximum of W_1^s is transformed into a minimum. Since W_1^s must be zero at the boundaries, a new maximum of W_1^s must be formed somewhere (point P in Fig. 5). Since the symmetric point is already occupied with the minimum of W_1^s, the new maximum must form on a less symmetric point. Therefore, it breaks the symmetry and is degenerate with a whole group of other maxima. The new stationary state (3.1) of the system is now given by one of these less symmetric maxima, i.e., a symmetry changing transition has occurred. This behaviour is well known for systems in thermal equilibrium undergoing a second order phase transition and concepts of second order phase transitions may, in fact, be applied to this problem. It should be noted, however, that most of the difficulties of phase transition theory can be avoided here, because they are due to the necessity of taking the thermodynamic limit of an infinite system. This limit has not to be taken for the examples we consider here. Therefore, the mean field theory of phase transitions, which disregards the singularities due to the thermodynamic limit, is particularly well suited for our cases. Its derivation in terms of pure symmetry arguments was given by Landau [17]. We apply his reasoning to determine W_1^s in the vicinity of $\{\lambda\}=\{\lambda_c\}$.

Let G be the symmetry group describing the symmetries of the branch of states with higher symmetry. Then the state $\{w^s(\lambda_c\})$ is an invariant of G. In the vicinity of the transition the states on the less symmetric branch differ little from $\{w^s(\{\lambda_c\})\}$ and we may put

$$\{w^s(\{\lambda\})\}=\{w^s(\{\lambda_c\})\}+\{\Delta w^s(\{\lambda\})\} \tag{3.21}$$

with small $\{\Delta w^s\}$. The potential $\phi^s(\{w\})$ can now be determined from the condition, that (3.21) gives its minima (cf. Eq. (3.8)). Since $\{\Delta w^s(\{\lambda\})\}$

is small we may expand ϕ^s in a power series of $\{\Delta w\} = \{w\} - \{w^s(\{\lambda_c\})\}$. Since ϕ^s is an invariant of G it can only depend on invariants which can be formed by powers and products of the variables $\{\Delta w\}$. There is no first order invariant of G besides $\{w^s(\{\lambda_c\})\}$. Hence, the power series starts with the second order invariants $F_\nu^{(2)}(\{\Delta w\})$, one invariant being connected with each irreducible representation ν of G. The invariants $F_\nu^{(2)}(\{\Delta w\})$ can all be chosen to be positive. This gives

$$\phi^s = \sum_\nu a_\nu F_\nu^{(2)}(\{\Delta w\}) + \cdots . \tag{3.22}$$

For $a_\nu > 0$ the minimum of ϕ^s is given by $\{\Delta w^s\} = 0$, and describes the symmetric branch. All $F_\nu^{(2)}$ are zero on this branch. A symmetry changing instability occurs, if at least one of the coefficients a_ν changes sign for $\{\lambda\} = \{\lambda_c\}$. The corresponding invariant $F_\nu^{(2)}$ will then have a non-zero value in the stationary state, and higher order terms in the expansion are required. The third order invariants have to vanish if $\{\Delta w^s(\{\lambda_c\})\}$ is to be a stable state and the 4th order terms have to be positive definite. The potential ϕ^s is then given by

$$\phi^s = aF^{(2)}(\{\Delta w\}) + \sum_\mu b_\mu F_\mu^{(4)}(\{\Delta w\}) . \tag{3.23}$$

In this expansion all second order invariants have been dropped, besides the one invariant $F^{(2)}$, whose coefficient a changes sign at the transition point. The other invariants describe fluctuations which are weak compared to the strong fluctuations arising from the transition. The latter are only limited by the 4th order terms in the expansion. For the same reason, only the fourth order invariants of the corresponding irreducible representation have to be taken into account. This limits the number of phenomenological coefficients a, b which have to be introduced. The expansion (3.23) may be simplified further by introducing the new variables

$$\eta^2 = F^{(2)}(\{\Delta w\}) , \tag{3.24}$$

$$\{\Delta \hat{w}\} = \eta^{-1}\{\Delta w\} . \tag{3.25}$$

Since the second order term in Eq. (3.23) depends on η only, the fluctuations in $\{\Delta \hat{w}\}$ are small, so that these variables can be replaced by the quantities which minimize ϕ^s under the constraint

$$F^{(2)}(\{\Delta \hat{w}\}) = 1 . \tag{3.26}$$

The remaining expression

$$\phi^s = a\eta^2 + b\eta^4 \tag{3.27}$$

with

$$a \sim (\lambda_c - \lambda); \qquad b > 0 \tag{3.28}$$

gives ϕ^s and the stationary distribution W_1^s as a function of the second order invariant (3.24) alone. Thus, in the vicinity of a symmetry changing instability the number of variables, on which the potential ϕ^s and the stationary distribution $W_1^s \sim \exp(-\phi^s)$ depend, is effectively reduced to 1. This will simplify the analysis of the dynamics considerably.

3.3. Dissipation-Fluctuation Theorem for Stationary Nonequilibrium States

The linear response of a system[7], described by Eq. (2.10), to an external perturbation can easily be calculated by adding a perturbation term on the right hand side of Eq. (2.10). We obtain

$$\partial W_1/\partial t = L W_1 + L_{ext} W_1 . \tag{3.29}$$

Here, L is the linear operator acting on W_1 on the right hand side of Eq. (2.10). It fulfills the relation

$$L W_1^s = 0 . \tag{3.30}$$

The operator L_{ext} describes an additional external perturbation. In general, it will take the form of a Poisson bracket with a perturbation Hamiltonian H_{ext}.

$$L_{ext} W_1^s = [H_{ext}, W_1^s] \equiv \left[\frac{\partial H_{ext}}{\partial u_i}\frac{\partial W_1^s}{\partial v_i} - \frac{\partial W_1^s}{\partial u_i}\frac{\partial H^{ext}}{\partial v_i}\right]. \tag{3.31}$$

In defining the Poisson bracket in Eq. (3.31) we have assumed that we can split the variables $\{w\}$ into pairs of generalized coordinates $\{u\}$ and momenta $\{v\}$. This is not a real restriction, since for each coordinate we may formally introduce a conjugate momentum, on which ϕ^s depends as a second order function. At the end of the calculations we may eliminate these variables by integrating over them. H_{ext} is then the Hamiltonian of the external perturbation which has the general form

$$H_{ext}(t) = -A_i(\{u\}, \{v\}) F_i(t) . \tag{3.32}$$

Here, $\{F(t)\}$ is a set of external forces coupled to the system by some functions $\{A(\{u\}, \{v\})\}$. By standard first order perturbation theory, we find the first order response ΔX of some function $X(\{u(t)\}, \{v(t)\})$ to

[7] For other calculations see [26] and [27]. The latter treatment is similar to the one given here.

the external force $F_i(\tau)$

$$\langle \Delta X(\{u(t)\}, \{v(t)\}) \rangle = \int_{-\infty}^{t} d\tau\, \phi_{X,i}(t-\tau)\, F_i(\tau)\,. \tag{3.33}$$

The well known result for the response function $\phi_{X,i}(\tau)$

$$\phi_{X,i}(\tau) = \langle [A_i(\{u(t)\}, \{v(t)\}), X(\{u(t+\tau)\}, \{v(t+\tau)\})] \rangle \tag{3.34}$$

is the average of a two-time Poisson bracket. Expressing W_1^s by ϕ^s we obtain

$$\phi_{X,i}(\tau) = -\langle X(\{u(t+\tau)\}, \{v(t+\tau)\})\, [\phi(\{u(t)\}, \{v(t)\}), A_i(\{u(t)\}, \{v(t)\})] \rangle \tag{3.35}$$

which is the two-time correlation function of the function X and a Poisson bracket. This result is similar to the result for thermal equilibrium systems. There, ϕ^s is replaced by the Hamiltonian H and the Poisson bracket reduces to a first order derivative in time. Apart from special cases, no general relation between ϕ^s and the evolution in time exists in stationary nonequilibrium systems. Hence, this last step cannot be performed in this general case. In the special case of systems which have the property of detailed balance in the stationary state, a further simplification is possible. These systems are considered in the next section.

4. Systems with Detailed Balance

In the discussion of the stationary distribution in the preceeding section we could make use of many considerations familiar from systems in thermal equilibrium. In general, this analogy does not hold for the dynamic behaviour. As indicated in 3.3, the stationary distribution contains only a little information about the dynamic behaviour of the system. The reason is, as we will see in this section, the lack of detailed balance in stationary nonequilibrium states. It is the presence of detailed balance in thermal equilibrium, which provides there the important link between statics and dynamics. Therefore, the special class of stationary nonequilibrium systems exhibiting detailed balance with respect to their relevant variables $\{w\}$ should show a close analogy to thermal systems, even with respect to their dynamic behaviour. The detailed balance of stationary nonequilibrium systems will not be complete and will not comprise all degrees of freedom, because of the action of external forces and fluxes. Fortunately, it is sufficient for our purposes to consider systems showing detailed balance with respect to the small number of variables $\{w\}$ which are used to describe the system. Detailed balance is discussed from a general point of view in [28]. Some implications for Markoffian processes were considered in [21]. Our analysis follows the recent papers [29, 30].

4.1. Microscopic Reversibility and Detailed Balance

In the following the transformation of the variables $\{w\}$ with time reversal is important. We define a new set

$$\{\tilde{w}\} = \{\varepsilon_1 w_1, \varepsilon_2 w_2, \dots, \varepsilon_n w_n\} \tag{4.1}$$

where $\varepsilon_i = -1(+1)$[8], if w_i does (does not) change sign if time is reversed. (The variables can always be chosen that either of these are true.) Similarly we consider the time reversal transformation of a set of externally determined parameters $\{\lambda\}$, on which the probability densities may depend, and define

$$\{\tilde{\lambda}\} = \{\nu_1 \lambda_1, \nu_2 \lambda_2, \dots, \nu_n \lambda_n\} \tag{4.2}$$

where $\nu_i = -1(+1)$, if λ_i does (does not) change sign if time is reversed. The property of microscopic reversibility may now be defined by the relation

$$W_2(\{w^{(2)}\}, t+\tau; \{w^{(1)}\}, t; \{\lambda\}) = W_2(\{\tilde{w}^{(2)}\}, t-\tau; \{\tilde{w}^{(1)}\}, t; \{\tilde{\lambda}\}) \tag{4.3}$$

where the dependence of the probability densities on the external parameters $\{\lambda\}$ has been made explicit. By specializing microscopic reversibility (4.3) for the stationary state we obtain the property of detailed balance

$$W_2^s(\{w^{(2)}\}, t+\tau; \{w^{(1)}\}, t; \{\lambda\}) = W_2^s(\{\tilde{w}^{(1)}\}, t+\tau; \{\tilde{w}^{(2)}\}, t; \{\tilde{\lambda}\}) . \tag{4.4}$$

Equation (4.4) expresses the following property of the stationary state: The number of transitions from $\{w^{(1)}\}$ at $t=t_1$ to $\{w^{(2)}\}$ at $t=t_2$ is equal to the number of transitions from $\{\tilde{w}^{(2)}\}$ at $t=t_1$ to $\{\tilde{w}^{(1)}\}$ at $t=t_2$. Therefore, apart from reversible motions, each pair of states $\{w^{(1)}\}$, $\{w^{(2)}\}$ is separately balanced in the stationary state. By using Eq. (2.3) we may rewrite Eq. (4.4) in the form

$$\begin{aligned} &P(\{w^{(2)}\} | \{w^{(1)}\}; \tau; \{\lambda\})\, W_1^s(\{w^{(1)}\}; \{\lambda\}) \\ &\qquad = P(\{\tilde{w}^{(1)}\} | \{\tilde{w}^{(2)}\}; \tau; \{\tilde{\lambda}\})\, W_1^s(\{\tilde{w}^{(2)}\}; \{\tilde{\lambda}\}) . \end{aligned} \tag{4.5}$$

Integrating Eq. (4.5) over $\{w^{(2)}\}$ we obtain a symmetry condition for $W_1^s(\{w\})$

$$W_1^s(\{w\}, \{\lambda\}) = W_1^s(\{\tilde{w}\}, \{\tilde{\lambda}\}) . \tag{4.6}$$

For systems in thermal equilibrium Eq. (4.5) can be derived from the time reversal invariance of the microscopic equations of motion. This derivation is no longer possible for systems in stationary nonequilibrium states, since external forces and fluxes will destroy detailed balance. The station-

[8] In all formulas containing ε_i and ν_i no summation over repeated indices is implied.

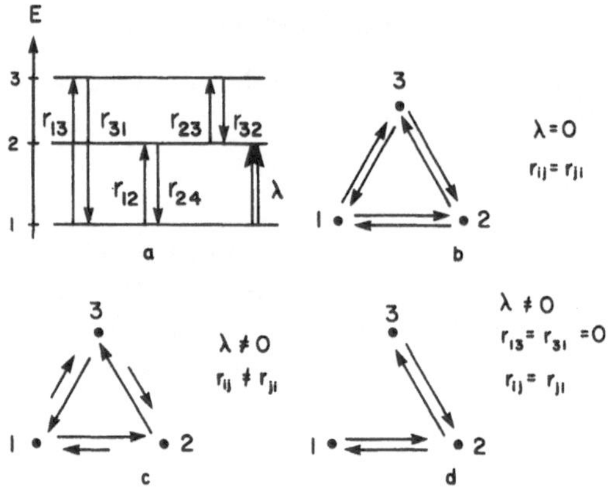

Fig. 6a – d. Stationary states with and without detailed balance for a 3-level atom. a Energy levels with transitions rates r_{ij} and pump rate λ. b Equilibrium ($\lambda = 0$) with detailed balance. c Stationary nonequilibrium state ($\lambda \neq 0$) without detailed balance. d Stationary non-equilibrium state ($\lambda \neq 0$) with detailed balance for $r_{13} = r_{31} = 0$.

ary distribution will then be maintained by cyclic sequences of transitions between more than two states [28]. The example of an externally pumped three-level atom, shown in Fig. 6, has been discussed in the literature [28, 31]. This example makes it obvious, that, detailed balance in a stationary nonequilibrium system will be present, if each pair of states is connected by only one sequence of allowed transitions. In Fig. 7, we give

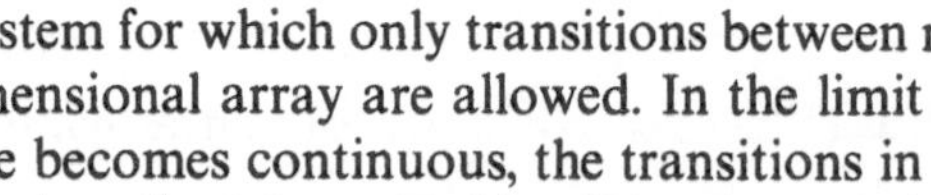

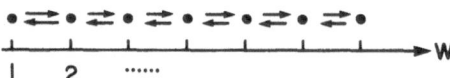

Fig. 7. Detailed balance in a one-dimensional array of states with transitions between neighbouring states.

as an example, a system for which only transitions between neighbouring states in a one-dimensional array are allowed. In the limit in which the configuration space becomes continuous, the transitions in this example would have to be described by a Fokker-Planck equation in a one-dimensional configuration space. If the transitions have to vanish at the boundaries of the configuration space, it is obvious from Fig. 7 that detailed balance has to be present in the stationary state. In all cases, in which the configuration space of the system has more than one dimension (cf. Fig. 8), each pair of states is connected by many different sequences of allowed transitions, even if only transitions between neighbouring states in configuration space are allowed. In these cases, detailed balance is guaranteed, if symmetry demands that the transition rate from one state to some other state is equal for all possible sequences of intermediate states. E.g., if the external forces acting on the system represented by Fig. 8 can only cause transitions between different states in radial direction, and if a rotation of phase space leaves the system properties unchanged, the boundary conditions are still sufficient to guarantee the presence of detailed balance.

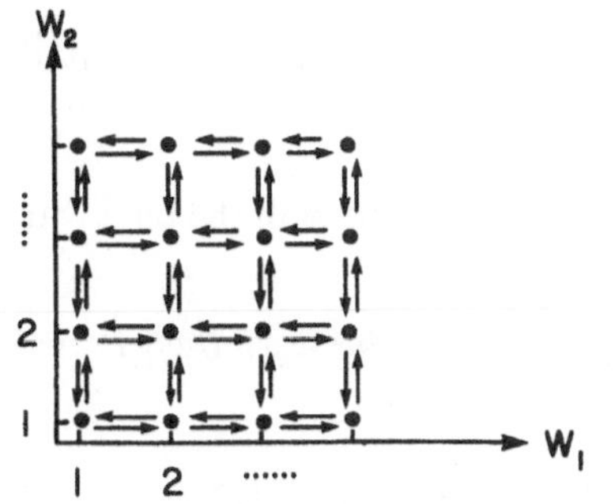

Fig. 8. Detailed balance in a two-dimensional array of states with transitions between neighbouring states

Detailed balance due to symmetry is of special importance for stationary nonequilibrium systems in the vicinity of a symmetry changing instability. For such systems an expression for the potential ϕ^s was obtained in Section 3.2. This expression can be inserted into Eq. (2.24) in order to obtain an equation of motion. If the external forces acting on the system enter this equation of motion only by the derivative $\partial\phi^s/\partial w_k$ and not by $\{r^s\}$, detailed balance has to be present in the stationary state because of symmetry, for the following reason. The external forces determine the coefficient a in Eq. (3.27) and are thus coupled to the system only by a second order invariant; this coupling can only cause transitions between states having different values of the second order invariant; the boundary conditions are sufficient to guarantee detailed balance with respect to these transitions. Transitions between states without change of the second order invariant are not influenced by the external forces and, hence, are in detailed balance as well. This general mechanism explains why many of the stationary nonequilibrium systems which are considered in part B have the property of detailed balance.

4.2. The Potential Conditions

In this section, we derive the conditions which have to be satisfied by the drift and diffusion coefficients of Eq. (2.10), in order to guarantee detailed balance in the stationary state [29]. To this end we solve Eq. (4.5) for $P(\{w\}|\{w'\};\tau;\{\lambda\})$ and insert the resulting expression into Eq. (2.10), which P must satisfy. The equation for $P(\{\tilde{w}'\}|\{\tilde{w}\};\tau;\{\tilde{\lambda}\})$, which we obtain in this way, is simplified by using the time independent equation of motion for the stationary distribution W_1^s. It takes the form

$$\begin{aligned}\{\partial/\partial\tau+[K_i(\{w\};\{\lambda\})-\partial K_{ik}(\{w\};\{\lambda\})/\partial w_k\\ +K_{ik}(\{w\};\{\lambda\})\,\partial\phi^s(\{w\};\{\lambda\})/\partial w_k-\tfrac{1}{2}K_{ik}(\{w\};\{\lambda\}\,\partial/\partial w_k]\,\partial/\partial w_i\}\\ \cdot P(\{\tilde{w}'\}|\{\tilde{w}\};\tau;\{\tilde{\lambda}\})=0\,.\end{aligned} \tag{4.7}$$

This equation is now compared with the backward equation (2.12), which we may rewrite in the form

$$\{\partial/\partial\tau-(\tilde{K}_i(\{w\})+\tfrac{1}{2}\tilde{K}_{ik}(\{w\})\,\partial/\partial w_k)\,\partial/\partial w_i\}\,P(\{\tilde{w}'\}|\{\tilde{w}\};\tau;\{\tilde{\lambda}\})=0 \tag{4.8}$$

by substituting

$$\{w'\}\to\{\tilde{w}\};\qquad\{w\}\to\{\tilde{w}'\};\qquad\{\lambda\}\to\{\tilde{\lambda}\} \tag{4.9}$$

and introducing the notation

$$\begin{aligned}\tilde{K}_i\,(\{w\};\{\lambda\})&=\varepsilon_i K_i(\{\tilde{w}\};\{\tilde{\lambda}\})\\ \tilde{K}_{ik}(\{w\};\{\lambda\})&=\varepsilon_i\varepsilon_k K_{ik}(\{\tilde{w}\};\{\tilde{\lambda}\})\,.\end{aligned} \tag{4.10}$$

Eliminating the time derivative from Eqs. (4.7), (4.8) we obtain the identity

$$\begin{aligned}0=\{(\partial K_{ik}/\partial w_k-K_i-\tilde{K}_i-K_{ik}\,\partial\phi^s/\partial w_k)+\tfrac{1}{2}(K_{ik}-\tilde{K}_{ik})\,\partial/\partial w_k\}\\ \cdot\partial P(\{\tilde{w}'\}\,|\{\tilde{w}\}\,;\tau\,;\{\tilde{\lambda}\})/\partial w_i\end{aligned} \tag{4.11}$$

which holds for all times. All quantities in the curly brackets are functions of $\{w\}$ and $\{\lambda\}$. For $\tau=0$, P is a δ-function according to the initial condition (2.7). Multiplying Eq. (4.11) by an arbitrary function $F(\{w'\})$ and integrating over $\{w'\}$ for $\tau=0$, we obtain an identity, which contains terms linear in the first and second order derivatives of F. Since F and all its derivatives are arbitrary, the coefficients of all terms must vanish separately. This yields the potential conditions

$$K_{ik}(\{w\};\{\lambda\})=\varepsilon_i\varepsilon_k K_{ik}(\{\tilde{w}\};\{\tilde{\lambda}\}) \tag{4.12}$$

and

$$D_i-\tfrac{1}{2}\partial K_{ik}/\partial w_k=-\tfrac{1}{2}K_{ik}\,\partial\phi^s/\partial w_k\,. \tag{4.13}$$

In Eq. (4.13) we introduced the "irreversible drift"

$$D_i(\{w\};\{\lambda\})=\tfrac{1}{2}(K_i(\{w\};\{\lambda\})+\varepsilon_i K_i(\{\tilde{w}\};\{\tilde{\lambda}\})) \tag{4.14}$$

which transforms like w_i if time is reversed. Eqs. (4.12), (4.13) can be combined with Eq. (2.10) to yield

$$\partial J_i W_1^s/\partial w_i = 0\,. \tag{4.15}$$

Here we introduced the "reversible drift"

$$J_i(\{w\};\{\lambda\}) = \tfrac{1}{2}(K_i(\{w\};\{\lambda\}) - \varepsilon_i K_i(\{\tilde{w}\};\{\tilde{\lambda}\})) \tag{4.16}$$

which transforms like $\dot{w}_i$ if time is reversed. The drift coefficient K_i is given by the sum

$$K_i = D_i + J_i\,. \tag{4.17}$$

So far we have shown that the potential conditions (4.12), (4.13) are necessary for the compatibility of Eq. (2.10) with the condition of detailed balance (4.5). In order to show that they are also sufficient, we derive now the symmetry relation (4.5) from the conditions (4.12), (4.13) by assuming that the Fokker-Planck equation and its adjoint (2.12) hold. Since Eqs. (2.10), (4.12), (4.13) hold, the identity (4.11) is certainly fulfilled. Using Eq. (2.12) in its form (4.8), we may work from Eq. (4.11) backwards and obtain the Fokker-Planck equation (2.10) for the quantity $P(\{\tilde{w}'\}|\{\tilde{w}\};\tau;\{\tilde{\lambda}\})\, W_1^s(\{\tilde{w}\},\{\tilde{\lambda}\})$.

The drift and diffusion coefficients of this Fokker-Planck equation depend on $\{w\}$, $\{\lambda\}$. By assumption, the same equation with the same initial and boundary conditions holds for the quantity $P(\{w\}|\{w'\};\tau;\{\lambda\})$. In as much as the Green's function for the Fokker-Planck equation with natural boundary conditions is unique apart from a normalization constant N, we may equate the two quantities

$$P\{\tilde{w}'\}|\{\tilde{w}\};\tau;\{\tilde{\lambda}\})\, W_1^s(\{\tilde{w}\};\{\tilde{\lambda}\}) = N(\{w'\};\{\lambda\})\, P(\{w\}|\{w'\};\tau;\{\lambda\})\,. \tag{4.18}$$

Integrating over $\{w\}$ we obtain

$$W_1^s(\{\tilde{w}'\};\{\tilde{\lambda}\}) = N(\{w'\};\{\lambda\}) \tag{4.19}$$

whereby Eq. (4.18) is reduced to the relation (4.5). Hence, the potential conditions (4.12), (4.13) and the detailed balance condition (4.5) are equivalent for all systems which are described by Eq. (2.10) and the backward equation (2.12).

The potential conditions (4.12), (4.13) impose severe restrictions on the coefficients $\{D\}$, $\{J\}$, and K_{ik} of the Fokker-Planck equation (2.10). From Eq. (4.13) we obtain by differentiating

$$\partial/\partial w_j\,\{K_{il}^{-1}(\partial K_{lk}/\partial w_k - 2D_l)\} = \partial/\partial w_i\,\{K_{jl}^{-1}(\partial K_{lk}/\partial w_k - 2D_l)\} \tag{4.20}$$

where the existence of K_{il}^{-1} is assumed. From Eq. (4.15) we obtain, by eliminating W_1^s with the help of Eq. (4.13),

$$\partial J_i/\partial w_i - J_i K_{il}^{-1}(\partial K_{lk}/\partial w_k - 2D_l) = 0\,. \tag{4.21}$$

Special cases of these conditions have already been discussed in the literature on stationary nonequilibrium systems [20, 21]. Their practical importance in laser theory has also been recognized [32]. For systems in thermal equilibrium detailed balance is a general property. Hence, the potential conditions have always to be satisfied in equilibrium theory. In fact, a look at the general Fokker-Planck equations, derived for systems near thermal equilibrium, confirms that the potential conditions are satisfied by the drift and diffusion coefficients of these equations [33, 34].

4.3. Consequences of the Potential Conditions

The meaning of Eqs. (4.12) – (4.17) is analyzed best by a comparison with the more general Eqs. (2.20), (2.22). First of all, we note that $\{J\}$, defined by Eq. (4.16), is the drift velocity in the stationary state

$$J_i = r_i^s\,. \tag{4.22}$$

Since J_i transforms like $\dot{w}_i$ (if time is reversed), J_i describes all reversible drift processes. The remaining part of K_i is given by D_i and describes all irreversible drift processes. We find, therefore, that the general decomposition of the total drift into two parts, as introduced in Eq. (2.22), coincides, in the presence of detailed balance, with the general decomposition of the total drift into a reversible and an irreversible part. The general result of the preceeding section can now be formulated as follows:

Systems, described by Eqs. (2.10), (2.12) are in detailed balance in their stationary state, if and only if the probability current in the stationary state is the reversible part of the drift. We note that, in detailed balance, cyclic probability currents are not forbidden altogether; only irreversible probability currents are not allowed.

By introducing the potential conditions (4.12), (4.13), into the Langevin Eqs. (2.24) we obtain

$$\dot{w}_i = J_i - \tfrac{1}{2}K_{ik}\,\partial\phi^s/\partial w_k + g_{ij}(\tfrac{1}{2}\partial g_{kj}/\partial w_k + \xi_j)\,. \tag{4.23}$$

These equations show the close analogy which exists between systems near equilibrium and systems near stationary nonequilibrium states [30]. Eq. (4.13) is the analogue of the linear, phenomenological relations of irreversible thermodynamics [13] between the "generalized forces", represented by the derivatives of ϕ^s, and the "generalized irreversible fluxes", represented by the irreversible drift. The potential ϕ^s plays the role of a thermodynamic potential, both, in its static and its dynamic aspects. The diffusion coefficients K_{ik} are the analogue of the coefficients in the linear relations between fluxes and forces. Eqs. (4.12) are the analogue of the Onsager-Casimir symmetry relations [35, 36] for these coefficients.

The potential conditions (4.12), (4.13) have considerable practical importance, since Eq. (4.13) gives n first integrals of the time independent Fokker-Planck equation for W_1^s. These first integrals may be used in two different ways:

i) It is possible to determine the stationary distribution $W_1^s \sim \exp(-\phi^s)$ from Eq. (4.13) by the line integral

$$\phi^s = -2\int K_{ik}^{-1} [D_k - \tfrac{1}{2} \partial K_{kj}/\partial w_j] \, dw_i \tag{4.24}$$

if the drift and diffusion coefficients are known. Eq. (4.13) will be used in this manner in Section 8.

ii) It is possible to determine the irreversible drift $\{D\}$, if the diffusion matrix K_{ik} and the stationary distribution W_1^s are known. In this way it is possible to extract information on the dynamics of the system from the stationary distribution. This procedure is of importance in all cases in which symmetry arguments, like those of Section 3.2, are sufficient to obtain the stationary distribution and the diffusion matrix. We will use it in the applications of Sections 6 and 7.

In all cases of vanishing reversible drift, $J_i = 0$, the quantity ϕ^s and the diffusion coefficients determine both the dynamics and the stationary distribution. Eq. (4.13) is then a somewhat disguised form of the fluctuation dissipation theorem, since it gives the dissipative drift in terms of the fluctuations. It can be converted to the more usual form of the fluctuation dissipation theorem by considering the linear response of the variable w_i to an external force, driving the variable w_j. The response is given by Eq. (3.35), if we take A_j to be the momentum which is canonically conjugate to w_j. The response function is then given by

$$\phi_{ij}(\tau) = -\langle w_i(\tau) \, \partial\phi^s/\partial w_j \rangle \,. \tag{4.25}$$

By using Eq. (4.13) we obtain

$$\phi_{ij}(\tau) = 2\langle K_{jk}^{-1} (D_k - \tfrac{1}{2} \partial K_{kl}/\partial w_l) \, w_i(\tau) \rangle \,. \tag{4.26}$$

If we assume that K_{ij} is independent of $\{w\}$ and use Eq. (4.23), we obtain the more familiar form

$$\phi_{ij}(\tau) = -2K_{jk}^{-1} \, \partial \langle w_i(t) \, w_k(t-\tau) \rangle / \partial\tau \,. \tag{4.27}$$

In deriving Eq. (4.27) from (4.26) and (4.23) we made use of the fact that the fluctuating forces $g_{ij} \xi_j(t)$ in Eq. (4.23) give no contribution in Eq. (4.26), since there, τ is always positive and all correlations vanish. The results (4.26), (4.27) coincide with results obtained recently by Agarwal [27].

In all cases of vanishing irreversible drift $D_i=0$, Eq. (4.13) yields $K_{ik}=0$. In this case, the potential ϕ^s cannot be determined from Eq. (4.13). It rather has to be determined from Eq. (4.15) in terms of the reversible drift $\{J\}$. In most cases the latter can be derived from a Hamiltonian H by splitting the variables $\{w\}$ into pairs of canonically conjugate coordinates $\{u\}$ and momenta $\{v\}$ and putting

$$J_i^{(u)} = \partial H/\partial v_i; \qquad J_i^{(v)} = -\partial H/\partial u_i . \tag{4.28}$$

In this case, our theory is formally reduced to equilibrium theory. The stationary distribution can be taken to be the canonical distribution

$$W_1^s \sim \exp(-H/T) \tag{4.29}$$

where T is some fluctuation temperature in energy units. The fluctuation dissipation theorem (3.35) reduces to its equilibrium form. $\phi^s = H/T$ determines both the dynamics and the stationary distribution completely.

B. Application to Optics

5. Applicability of the Theory to Optical Instabilities

In the second part of this paper we consider threshold phenomena in nonlinear optics. Thresholds in laser physics and nonlinear optics mark the onset of instability of certain modes of the light field. In this section we consider some common features of these instabilities and discuss the relevance of the general part A for laser physics and nonlinear optics. In Section 5.1 we consider the validity of the basic assumptions and give a review of the quantities which connect the theory and photocount experiments. In Section 5.2 we give an outline of the microscopic theory of fluctuations in lasers and nonlinear optics. This outline is necessary, since we will make use of the microscopic theory in Section 8. Furthermore, the results of the phenomenological theory in Sections 6 and 7 will frequently be compared with results of the microscopic theory. In Section 5.3 we discuss the general analogy between instabilities in nonlinear optics and second order phase transitions. These analogies are a special case of the general connections between symmetry changing instabilities of stationary nonequilibrium states and second order phase transitions. The limits of this analogy, which are due to the geometry of optical systems, are also discussed.

5.1. Validity of the Assumptions; the Observables

Before applying the considerations of part A to optical examples, we have to check the validity of the basic assumptions and have to find the observables of photo-count experiments.

a) The Assumptions

i) *Stationarity* implies the time independence of all external influences on the system, on the adopted time scale of description. Hence, all parameters which characterize a given optical device, like temperature, distances and angles between mirrors, intensity and mode pattern of pump sources, have to be stabilized on that time scale. This stabilization presents experimental difficulties, which could be overcome for single mode lasers [10]. For most other optical oscillators stabilization is more difficult, either because their mode selection mechanisms are less efficient (e.g. parametric oscillators), or because they depend more critically on properties of the pump (e.g. Raman Stokes oscillator). Nevertheless, recent technological progress [37] should make a stabilization of other oscillators, like parametric oscillators, over sufficiently long time intervals, possible.

ii) The assumption of the validity of a *Fokker-Planck equation* can be split into the Markoff assumption and the diffusion assumption. In nonlinear optics, a Markoff description is usually provided by the amplitudes of the optical modes and the variables of the medium which account for the nonlinear interaction (cf. 5.2). In our phenomenological theory, the variables, which are used to describe the system, are the amplitudes of the unstable modes alone. Whether this restriction of the number of variables is justified or not depends on whether the system is sufficiently close to the instability, since the lifetime of the fluctuations of the unstable mode amplitude becomes large in the vicinity of the instability. The necessary number of variables also depends on the time scale of observation, which is determined by the rise time of the photo diode ($\sim 10^{-9}$ sec) of the detector. Theoretical [38] and experimental [39] investigations of a possibly non-Markoffian behaviour of the single mode laser amplitude on the *n* sec time scale have been made. Experimentally, non-Markoffian effects have not been observed. Hence, the Markoff assumption seems to be well justified, at least for single mode instabilities.

The diffusion approximation can generally be justified for all optical modes with sufficiently high intensities. Fluctuations in optical modes are due to processes which involve the creation and annihilation of single light quanta. Jumps of the quantum number by ± 1 can be approximated by a continuous diffusion, if the total quantum number is sufficiently large.

Together with the Fokker-Planck equation, we introduced natural boundary conditions in part A. Their physical basis in nonlinear optics is the condition, that infinite field amplitudes occur with probability zero.

iii) In most optical applications we will restrict ourselves to systems with *detailed balance*. This assumption can be justified on general grounds only for special cases, most importantly the single mode laser treated in 6.1. In all other cases, it implies a restriction to special systems, whose parameters are chosen in such a way, that detailed balance is guaranteed. The potential conditions (4.12), (4.13) are a convenient tool to decide whether a system is in detailed balance or not.

b) The Observables

In most experiments of laser physics and nonlinear optics the interesting observables are the intensities of the light modes. Furthermore, the stability of the state of the system, i.e. the reproducibility of the results, is of interest. Theoretically, this information is provided by the description which neglects fluctuations, i.e. by the set of Eqs. (3.4). As was shown in Section 3, the symmetry changing instabilities have the most drastic effects on this level of description. They manifest themselves by a dramatic increase in the intensity of the instable mode, if treshold is passed [11]. In 3 it was also shown that the location of the minima of ϕ^s and the drift velocity $\{r^s\}$ determine the size of the stationary intensities and their stability.

In the last few years a growing number of experimentalists have been concerned with the statistical properties of the emitted light. Both the theoretical and the experimental details of their measurements have been the subject of many papers [8, 10, 18, 19]. Therefore, we restrict ourselves to a brief survey here. The quantity, which is on the basis of our phenomenological theory, is the stationary distribution of the mode amplitudes. It is closely connected with the most fundamental quantity for photo-count experiments, the stationary photo-count distribution $p(n, T, t)$. It gives the probability of counting n photoelectrons, which are generated by the light field in a photodiode within a given time interval T at time t. The photo-count distribution $p(n, T, t)$ depends on the statistical properties of the light field, since it is determined by averaging over a Poisson distribution

$$p(n, T, t) = \langle n!^{-1}\, \bar{n}(T, t)^n \exp(-\bar{n}(T, t))\rangle \tag{5.1}$$

whose mean value $\bar{n}$ is proportional to the average of the light intensity $I(t)$ [40]

$$\bar{n}(T, t) = \alpha \int_t^{t+T} I(t')\, dt'. \tag{5.2}$$

α gives a measure of the efficiency of the counting method. The average in (5.1) has, in general, to be taken with a probability density which is a functional of the intensity $I(t')$ for all times $t \leqq t' \leqq t+T$. However, if the interval T (which is determined by the rise time of the photodiode) is much shorter than the time scale on which $I(t)$ varies, Eq. (5.1) may be reduced to

$$p(n, T, t) = \int_0^\infty \mathrm{d}I\, n!^{-1} (\alpha I T)^n \, \mathrm{e}^{-\alpha I T}\, W_1^s(I, t)\,. \tag{5.3}$$

The measurement of $p(n, T, t)$ gives an indirect determination of W_1^s. W_1^s can also be characterized by its normalized moments $\langle I(t)^k \rangle / \langle I(t) \rangle^k$. They are given in terms of the normalized factorial moments $n^{(k)}$ of the photo-count distribution,

$$n^{(k)}(T, t) \equiv \langle n \rangle^{-k} \sum_n n!(n-k)!^{-1}\, p(n, T, t)\,, \tag{5.4}$$

by the relation

$$\langle I(t)^k \rangle / \langle I(t) \rangle^k = n^{(k)}(T, t)\,. \tag{5.5}$$

Usually, a comparison of the theoretical and experimental results for the first few moments is used, to fit the unknown parameters in W_1^s. Increasing the accuracy in the determination of the distribution $p(n, T, t)$ means to increase the number of known normalized factorial moments $n^{(k)}$. Thereby one increases the number of known normalized moments of W_1^s, and hence, the precision with which W_1^s is known. Therefore, photo-count experiments can test ϕ^s over the whole configuration space, whereas intensity measurements can only contain information on the (sharp) minima of ϕ^s.

Similar to single photo-count distributions one can measure joint photo-count distributions by determining the number of photoelectrons generated at different times. They provide an experimental method to determine the joint probability densities, introduced in Eq. (2.2). In most cases, however, one is content with the measurement of the lowest order moments of the joint distributions. This is done, e.g., in Hanbury-Brown Twiss experiments [41]. There, the photocurrents, produced in two or more photodetectors, placed in different space-time points (e.g. by beam splitters and electronic delay), are electronically multiplied and averaged over a time interval. In this way one is able to measure multi-time correlation functions, e.g. the autocorrelation function $\langle I(t+\tau)\, I(t) \rangle - \langle I(t) \rangle^2$, or cross-correlation functions like $\langle I_1(t+\tau)\, I_2(t) \rangle - \langle I_1(t) \rangle \langle I_2(t) \rangle$, if more than one mode of the electromagnetic field is excited. These quantities contain information about the dynamics of the system (e.g. relaxation times, fluctuation intensities). They can be calculated, either by the microscopic theory, which is reviewed in the next section, or by the phenomenological theory. While the microscopic theory is too involved to be applied to complicated problems, the phenomenological theory can also be applied to more complicated situations, but is then restricted to cases where detailed balance is present.

5.2. Basic Concepts of the Microscopic Theory

The general procedure of the microscopic theory is shown in a block diagram in Fig. 9. It was originally developed for the analysis of lasers (cf. [8]). Later, it was shown that the same procedure can be used in nonlinear optics. The starting point is a Hamiltonian which contains the following dynamical variables (operators):

i) The amplitudes of the electromagnetic field modes, described by boson creation and annihilation operators,

ii) the operators, describing the atoms of the medium, which obey anticommutator relations,

iii) a number of operators, describing incoherent pumping of the atoms of the field modes (e.g. in lasers), as well as dissipation and fluctuation due to the coupling to a number of thermal reservoirs, and

iv) c-number forces, describing external, coherent pumping (e.g. in parametric oscillators or Raman oscillators).

The approximations, which are usually made when the Hamiltonian is specified, are

i) the self consistent restriction of the field operators used, to the modes of the electromagnetic field which are strongly excited in the particular process under consideration,[9]

ii) the neglect of all interactions between the elementary excitations of the medium ("atoms"), except for the interaction mediated by the electromagnetic fields,

iii) restriction to resonant one-quantum processes for the interaction between light and matter (i.e. the dipole approximation and the rotating wave approximation).

Knowing the Hamiltonian one can write down the von Neumann equation of motion for the density operator of the whole system including the reservoirs. The main part of the theory consists now in a sequence of steps which simplify this equation, until it can be solved.

The first step is the elimination of the reservoir variables, which is most elegantly achieved by an application of Zwanzig's projector techniques, combined with a weak coupling approximation, and a Markoff assumption [43]. The latter implies that the correlation times

[9] The only exemption to this rule, known to the author, is the interesting work of Ernst and Stehle [42].

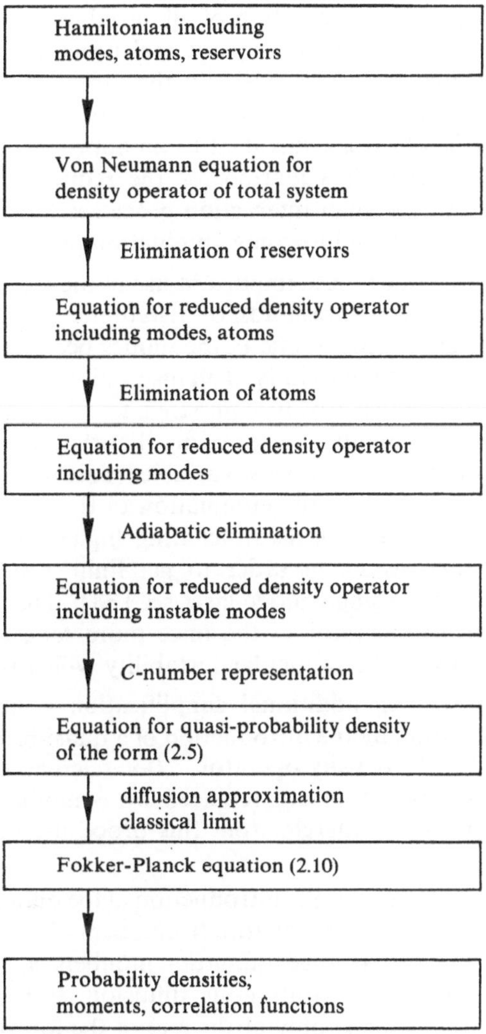

Fig. 9. Scheme of the microscopic theory

of the reservoirs are very short compared to all remaining time constants. As a result one obtains a "master equation" for the density operator in the reduced description, which contains the field modes and the variables of the medium. The reservoirs are now represented by a set of given external forces, described by time-independent parameters $\{\lambda\}$ and a set of damping and diffusion constants. The latter are connected by some fluctuation-dissipation relations which depend on the various reservoir temperatures.

The next step is the elimination of all variables which do not participate in the interaction with resonant real processes, but rather with nonresonant virtual processes. Usually the atomic variables play this role in nonlinear optics. This elimination can be achieved by a method described in [44], which is equivalent to an approximate unitary transformation. The remaining equation for the reduced density operator then describes only resonant interaction processes, whose coupling constants are obtained by the foregoing elimination process.

In the next step one makes important use of the fact that fluctuations are most important near thresholds, or instabilities. At these instabilities the inverse relaxation time of one of the modes becomes very small and changes sign. Hence, in the vicinity of an instability, there exists a number of variables which move slowly compared to all remaining variables. The latter may be eliminated by assuming that they are in a conditional equilibrium with respect to the slow variables (adiabatic approximation). The procedure is similar to the elimination of the reservoirs. The only difference is the necessity of also including higher order terms in the weak coupling expansion, in order to get finite results at threshold (for the example of the single mode laser see [38]). The remaining equation for the density operator of the once more reduced system holds only in the vicinity of the particular instability which is considered.

In the next step an additional simplification is achieved without further approximation by the introduction of a quasi-probability density representation for the density operator [10] (for references cf. [8]). In this representation all operators are replaced by c-number variables. The equation, which finally emerges from this procedure has the structure of Eq. (2.5).

The final simplification is the introduction of the diffusion approximation. Fluctuations change the quantum numbers of the modes by ± 1. For modes with large average quantum numbers $\bar{n}$, the fluctuations may be represented by a continuous diffusion. It is important that this approximation is made only at the end of the foregoing procedure, since, at the beginning, weakly excited degrees of freedom are also contained in the Hamiltonian.

The same argument which justifies the diffusion approximation can be used to apply the correspondence principle and take the classical limit of the final equation of motion. In this limit, the quasi-probability density is reduced to an ordinary probability density, as introduced in 2.1. By the procedure outlined above, a Fokker-Planck equation of the form (2.10) is obtained, which now has to be solved. Although this is a classical equation, it still describes quantum effects, since the fluctua-

[10] This step could also be done before the elimination procedure.

tions have a pure quantum origin. The fluctuations represent the small but measurable effects produced by the spontaneous emission process, which is conjugate to the stimulated process giving rise to the instability.

The advantage of the microscopic theory is the possibility to derive the drift and diffusion coefficients from first principles. Its disadvantages are its complexity, which restricts its applicability to simple systems, and the necessity for the introduction of many different approximations. In fact, many results of the microscopic theory are completely independent of the special form of the initial Hamiltonian and are only due to the occurrence of a symmetry changing transition. This is the main message conveyed by the phenomenological theory. Some of the results, which are independent of the special form of the initial Hamiltonian, are discussed in the next section and compared with phase transitions.

5.3. Threshold Phenomena in Nonlinear Optics and Phase Transitions

This section is devoted to a comparison between phase transitions in equilibrium systems and threshold phenomena in nonlinear optics. Analogies of this kind have been pointed out previously for the laser [45, 46] on the basis of the microscopic theory. Here, we discuss these analogies from a phenomenological point of view. We restrict ourselves to systems with detailed balance. Then the formal analogies between both classes of phenomena are obvious from the considerations in Sections 3, 4. It is sufficient to note that ϕ^s plays the role of a thermodynamic potential, both, in the static and in the dynamic domain, and that ϕ^s was constructed in analogy to the Landau theory of second order phase transitions in Section 3.2. However, a discussion of the analogies in more physical terms seems to be useful in order to appreciate their extent and their limits.

In both cases the basic instability arises from two competing processes. A phase transition [11] is determined by the competition between the thermal motion and a collective motion. The latter is caused by the interaction between the microscopic degrees of freedom, which, in the mean field approximation, is replaced by a nonlinear interaction of the microscopic degrees of freedom with a fictitious mean field. The nonlinear interaction gives rise to a positive feedback into a collective mode of the system. If the collective motion dominates, the mode becomes unstable. Its amplitude grows to a finite value, which is the order parameter of the phase transition. Observable order parameters must have zero frequency, since modes with finite frequency necessarily

[11] A qualitative discussion of phase transitions, which is suitable for our purposes here, is given in [47].

dissipate energy. At zero frequency thermal fluctuations are the dominant noise source.

Optical instabilities are governed by a competition between loss and gain in certain modes. The gain is due to a nonlinear interaction of the atoms (or elementary excitations) of the medium with the electromagnetic field, which plays the role of a mean field. However, contrary to the latter, the electromagnetic field is not fictitious. The characteristic length of the interaction is much longer in optical system (~ 1 m) than it is in systems with fictitious mean fields, where it is a microscopic quantity.

The mode, which becomes unstable by a feedback mechanism similar to the one before, has a finite frequency. This is possible, since the energy dissipation in this mode can be compensated by a stationary energy flow into the system. Thermal fluctuations are unimportant at optical frequencies. Instead, spontaneous emission processes are the main source of fluctuations.

In both cases, the instable mode of the system, if quantized, has to be a boson mode, because otherwise no positive feedback into this mode would be possible.

An important difference between usual phase transitions and optical instabilities comes from the difference in spatial dimensions. Optical devices have in most cases a one-dimensional geometry, and even the lengths in this single dimension are usually short compared to the coherence length of the electromagnetic field. Thus, the analogy has to be restricted to one and zero dimensional systems. In 3-dimensional systems the coherence length of the order parameter fluctuations diverges at the critical point. In the case of a continuous broken symmetry, the order parameter fluctuations contain an undamped zero frequency mode (Goldstone mode), which displaces the order parameter around a fixed, stable value, which breaks the symmetry. In 1 and 0-dimensional systems the order parameter fluctuations contain a damped zero frequency mode (diffusion mode), which carries the order parameter through a whole set of values, thereby restoring the symmetry [12]. The latter phenomenon takes the form of a phase diffusion in nonlinear optics. Besides this diffusion mode, there occur also fluctuations in the absolute value of the order parameter. These fluctuations are known to show a drastic slowing down in the vicinity of the critical point, because of the close matching between thermal and collective motion near that point [48]. Slowing down is also found near optical instabilities, where it is due to a close matching between the loss and the gain. At threshold, the total loss rate is equal to the sum of the gain by induced emission and the spontaneous emission rate. The spontaneous emission rate is smaller than the induced emission rate by a factor $1/\bar{n}$, where $\bar{n}$ is the mean number of quanta in the mode. If $\bar{n}$ would be infinite, a complete matching between loss and gain would be achieved at threshold and the slowing down would be critical [49].

In general, the slowing down decreases the decay rate of order parameter fluctuations at threshold up to a small but finite value, which is proportional to $1/\bar{n}$.

6. Application to the Laser

The general theory of part A is applied to the analysis of fluctuations of lasers in various operation modes. The example of the single mode laser, treated in 6.1, exhibits, in the simplest way, the general features outlined in part A. As this example has also been studied most carefully by experiments, it has to be considered as a prototype for the more complex examples studied in the later sections. In Section 6.2 we apply our theory to multimode operation in cases where mode coupling is only due to the intensities of the various modes, and no phase coupling is present. The systems treated in the Sections 6.1 and 6.2 represent examples with detailed balance. In Section 6.3 the case of multimode operation with phase coupling by various mechanisms is considered. In the case of self-locking, detailed balance is, in general, not present, due to irreversible cyclic probability currents through states with different relative phase angles of the modes. These currents make the analysis much more difficult, and only the simplest cases have been considered. The same applies to examples where phase interaction is forced from outside. However, some models, which are discussed in the literature, have the detailed balance property. Therefore, they may be analyzed by our methods.

In Section 6.4 we consider a system with one spatial dimension, the light propagation in an infinite laser medium. Two different states of the system are considered:

i) We treat the state which is most similar to single mode operation, but includes spatial fluctuations of the mode amplitude. This example shows most clearly, that a complete analogy to one-dimensional systems with complex order parameters exists (e.g. one-dimensional superconductors). The microscopic derivation of these results [46] originally suggested the development of a phenomenological theory, based on symmetry.

ii) We treat the state in which a periodic sequence of short pulses travels in the medium. The phenomenological theory is applied in order to show, how fluctuations (i.e. spontaneous emission) destroy periodicity over long distances.

6.1. Single Mode Laser

We consider a single mode of the electromagnetic field in resonance with externally pumped two-level atoms. The microscopic theory of this problem has been completely worked out since 1964 [50, 8]. Nevertheless we include this case in our analysis, because it exhibits most clearly the basic line of reasoning.

The threshold of laser action marks an instability of a mode of the electromagnetic field, whose electric fieldstrength we may write as

$$E(x,t) = (\beta \exp - i\omega_0 t + \beta^* \exp i\omega_0 t)\, f(x)\,. \qquad (6.1)$$

In (6.1) ω_0 is the laser frequency, which we assume to coincide with the atomic frequency, $f(x)$ is the normalized spatial pattern of the laser mode (running or standing wave), and β is the complex mode amplitude. Our general set of variables $\{w\}$ is formed by

$$\{w\} = \{w_1, w_2\} = \{\mathrm{Re}\,\beta, \mathrm{Im}\,\beta\}\,. \qquad (6.2)$$

The time reversal transformation behaviour of w_1, w_2 may be derived from Eqs. (6.1), (6.2) by noting that the electric field strength remains invariant. Hence, we find

$$\tilde{w}_1 = w_1\,, \qquad \tilde{w}_2 = -w_2\,. \qquad (6.3)$$

The external force λ, which keeps the system sufficiently far from thermal equilibrium, is supplied by the mechanism which inverts the electronic population of the two atomic levels participating in laser action.

Now we determine the potential $\phi^s(w_1, w_2)$ from symmetry arguments. It has to be invariant against changes of the phase angle of the complex mode β. In the completely symmetric state we have $w_1^s = w_2^s = 0$. Hence the quantities $\{\Delta w^s(\{\lambda\})\}$ defined in Eq. (3.21) are given by $\{w_1^s, w_2^s\}$. The potential ϕ^s is now obtained as a power series in $\{w_1, w_2\}$ containing only invariants formed by these quantities. The only invariants up to fourth order, which can be formed by these quantities, are $(w_1^2 + w_2^2) = |\beta|^2$ and $(w_1^2 + w_2^2)^2$. Hence we obtain

$$\phi^s = -a(w_1^2 + w_2^2) + b(w_1^2 + w_2^2)^2\,. \qquad (6.4)$$

The coefficient of the second order invariant has to change sign at threshold. Hence, we may put

$$a = \alpha(\lambda - \lambda_c)\,, \qquad b > 0, \qquad \alpha > 0\,. \qquad (6.5)$$

λ_c is the threshold value of the pump parameter λ. Since ϕ^s must have a minimum at $w_1 = w_2 = 0$ for $\lambda < \lambda_c$, we have $\alpha > 0$. The forms of ϕ^s for $\lambda \lessgtr \lambda_c$ are shown in Figs. 4, 5 respectively. The results (6.4), (6.5) are in complete agreement with the results obtained from the microscopic theory [51, 19, 8]. They are of central importance for the photon statistics of the single mode laser and have been checked experimentally with great care [52, 10]. Complete agreement between theory and experiment has been obtained.

In the next step, we derive the equation of motion (2.10). We assume that the diffusion matrix can be taken as independent of w_1, w_2. Because of phase angle invariance its only possible form is then

$$K_{ik} = \begin{pmatrix} q & 0 \\ 0 & q \end{pmatrix} \tag{6.6}$$

where q is another phenomenological constant. By applying Eq. (2.22) we obtain

$$K_i = -\tfrac{1}{2} q\, \partial \phi^s / \partial w_i + r_i^s \tag{6.7}$$

where r_i^s has to satisfy the equation

$$\partial r_i^s \exp(-\phi^s)/\partial w_i = 0\,. \tag{6.8}$$

Since the first term in Eq. (6.7) is a power series in $\{w\}$ we may also expand r_i^s as a power series. Observing phase angle invariance and the condition (6.8), we obtain

$$r_{1,2}^s = (a' - 2b' \, |\beta|^2)\, w_{2,1} \tag{6.9}$$

where a', b' are two real constants. Eq. (6.9) shows, that $\{r^s\}$ transforms like $\{\dot{w}\}$ and, hence, is a reversible drift. On the other hand $\{q\, \partial \phi^s / \partial w\}$ transforms like $\{w\}$ and is the irreversible part of $\{K\}$. As was proven in section (4.3) this is equivalent to the condition of detailed balance.

Since the reversible drift $J_{1,2} = r_{1,2}^s$, as given by Eq. (6.9), changes only the phase angle of β and leaves $|\beta|^2$ unchanged, it describes detuning effects. Since we assumed exact resonance between the atomic transition and the mode β of the field, we may put $a' = b' = 0$. The complete Fokker-Planck equation (2.10) now reads

$$\begin{aligned} \partial P / \partial \tau = & -(\partial/\partial w_1\, q(a - 2b|\beta|^2)\, w_1 P) - (\partial/\partial w_2\, q(a - 2b|\beta|^2)\, w_2 P) \\ & + \tfrac{1}{2} q (\partial^2 P / \partial w_1^2 + \partial^2 P / \partial w_2^2)\,. \end{aligned} \tag{6.10}$$

This result is again in complete agreement with the result of the microscopic theory [51] and with all experimental data obtained so far. The phenomenological parameters q, a, b are determined experimentally as follows [19]: qa as a function λ is obtained by fitting the experimental and theoretical results for the dimensionless quantity

$$\langle |\beta|^4 \rangle / |\langle \beta \rangle|^2 = \langle n^2 \rangle - \langle n \rangle / \langle n \rangle^2 \tag{6.11}$$

for different values of λ.

The right hand side of Eq. (6.11) contains moments of the photocount distribution (cf. Eqs. (5.4), (5.5)) which are accessible experimentally. The values of q and b are determined by measuring the average number of photons, $\langle|\beta|_c^2\rangle$, and the linewidth of the intensity fluctuations, $1/\tau_{I_c}$, at threshold. These quantities determine b and q by the relations [53]

$$\langle|\beta|_c^2\rangle = (\pi b)^{-1/2}; \qquad 1/\tau_{I_c} = q \cdot \sqrt{b} \cdot 5.854. \tag{6.12}$$

For a detailed presentation of the results obtained by the evaluation of Eq. (6.10), we refer to [53, 19, 8]. In Fig. 10 we show the results for the

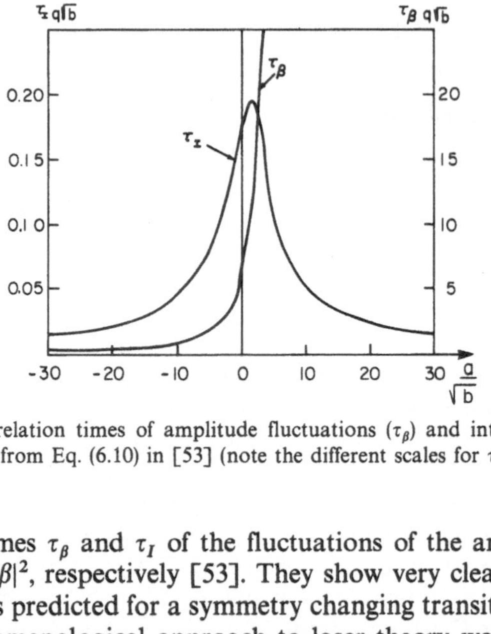

Fig. 10. The correlation times of amplitude fluctuations (τ_β) and intensity fluctuations (τ_I) as calculated from Eq. (6.10) in [53] (note the different scales for τ_β and τ_I)

correlation times τ_β and τ_I of the fluctuations of the amplitude β and the intensity $|\beta|^2$, respectively [53]. They show very clearly the slowing down which is predicted for a symmetry changing transition.

The phenomenological approach to laser theory was recently used by Grossman and Richter [54] – [56] to analyze the dynamics of lasers by a method which circumvents the use of Fokker-Planck equations. Their procedure runs as follows [54]:

i) The potential (6.4) is extended to include a "kinetic energy" term $\sim (\dot{w}_1^2 + \dot{w}_2^2) = |\dot{\beta}|^2$

$$\phi^s = d|\dot{\beta}|^2 - a|\beta|^2 + b|\beta|^4. \tag{6.13}$$

The new constant d has to be positive for normalization.

ii) The expression (6.13) is used as a Hamiltonian to generate equations of motion for the amplitude β. These equations are then modified by adding phenomenological damping terms.

iii) In order to determine correlation functions of the form $\langle F_1(\beta(t)) \cdot F_2(\beta(t_0))\rangle$, the equation of motion for $F_1(\beta(t))$ is solved with the initial condition $\beta(t_0)=\beta_0$. The result is multiplied by $F_2(\beta_0)$ and averaged over β_0 with the distribution $W_1^s \sim \exp(-\phi^s)$.

The same procedure is well known for systems in thermal equilibrium [17]. The steps ii) and iii) amount to a replacement of the Fokker-Planck equation by a suitable simplified set of moment equations. Besides this simplification, the most important difference to our procedure seems to be the fact that, in the treatment [54, 56], the main motion of the system is derived from a Hamiltonian and describes reversible processes whereas in our formulation the whole motion (besides detuning) was described by irreversible processes. Nevertheless, this procedure is equivalent to ours, apart from the additional approximations which are introduced by using simplified moment equations instead of the Fokker-Planck equation. We show this by deriving a Fokker-Planck equation from the potential (6.13) in the same way as before. Our new set of variables is now

$$\{w\}=\{w_1, w_2, w_3, w_4\}=\{\mathrm{Re}\,\beta, \mathrm{Im}\,\beta, \mathrm{Re}\,\dot{\beta}, \mathrm{Im}\,\dot{\beta}\}\,. \tag{6.14}$$

They obey the equations

$$\dot{w}_1=w_3\,, \qquad \dot{w}_2=w_4 \tag{6.15}$$

from which we obtain the drift and diffusion coefficients

$$K_1=w_3\,, \qquad K_2=w_4\,, \qquad K_{1i}=0; \qquad K_{2i}=0\,. \tag{6.16}$$

The diffusion matrix is taken to be constant and to preserve phase angle invariance. Then, it must take the form

$$K_{ik}=\begin{pmatrix} 0 & 0 & 0 & 0 \\ 0 & 0 & 0 & 0 \\ 0 & 0 & q_2 & 0 \\ 0 & 0 & 0 & q_2 \end{pmatrix}. \tag{6.17}$$

For the drift coefficients $K_{3,4}$ we obtain from Eq. (2.22)

$$K_{3,4}=-\tfrac{1}{2}q_2\,\partial\phi^s/\partial w_{3,4}+r^s_{3,4}\,. \tag{6.18}$$

From Eq. (2.20) we obtain

$$\partial r_i^s \exp(-\phi^s)/\partial w_i=0\,. \tag{6.19}$$

If we determine $r^s_{3,4}$ by a power series in $\{w\}$ which contains $w_{3,4}$ to the first order and $w_{1,2}$ to the third order, (the accuracy being determined by the accuracy of ϕ^s), and determine the coefficients of this expansion by Eq. (6.19), we obtain

$$r_3^s = \Delta\omega\, w_4 - \tfrac{1}{2} d^{-1}\, \partial\phi^s/\partial w_1 , \tag{6.20}$$

$$r_4^s = -\Delta\omega\, w_3 - \tfrac{1}{2} d^{-1}\, \partial\phi^s/\partial w_2 , \tag{6.21}$$

where $\Delta\omega$ is a real constant describing detuning effects. Neglecting detuning we take $\Delta\omega = 0$. Eqs. (6.15), (6.20), (6.21) show that $\{r^s\}$ transforms like a reversible drift. Therefore, detailed balance is present (cf. 4.3). The same equations show, that the potential (6.13), in fact, acts as a Hamiltonian for generating the reversible drift $\{J\} = \{r^s\}$. The irreversible drift consists of a linear phenomenological damping term. From Eq. (6.21) we obtain the Fokker-Planck equation

$$\begin{aligned} \frac{\partial P}{\partial t} = \Big\{ & -\frac{\partial}{\partial w_1} w_3 - \frac{\partial}{\partial w_2} w_4 - \frac{\partial}{\partial w_3} (w_1 d^{-1}(a - 2b|\beta|^2) - q_1\, d w_3) \\ & - \frac{\partial}{\partial w_4} (w_2 d^{-1}(a - 2b|\beta|^2) - q_2 d w_4) + \tfrac{1}{2} q_2 \left(\frac{\partial^2}{\partial w_3^2} + \frac{\partial^2}{\partial w_4^2} \right) \Big\} P . \end{aligned} \tag{6.22}$$

Its stationary solution is, of course, given by Eq. (6.13) with $W_1^s \sim \exp(-\phi^s)$. A method, which allows us to find the time dependent solutions of Eq. (6.22) if the solutions of Eq. (6.10) are known, is described in [20]. Eq. (6.10) is, of course, obtained from Eq. (6.22), if w_3, w_4 are eliminated as rapidly relaxing variables by an adiabatic approximation. The parameter q of Eq. (6.10) is then expressed in terms of the two parameters q_2, d of Eq. (6.22) by $q = (q_2 d^2)^{-1}$.

6.2. Multimode Laser with Random Phases

We now generalize the considerations of the preceding section to include the case of an arbitrary number of simultaneously excited modes. We assume that the mode amplitudes vary much more slowly in time than the atomic variables of the laser medium, whose characteristic times are given by the pumping time and the atomic relaxation times. Thus, the variables $\{w\}$ are the complex mode amplitudes β_ν. Furthermore, we assume that the laser operates in a region where the phases of all modes are independent from each other. Experimentally, this is a well known operation region.

The potential ϕ^s is given by the expansion [12]

$$\phi^s = -\sum_\nu a_\nu |\beta_\nu|^2 + \sum_{\nu,\nu'} b_{\nu\nu'} |\beta_\nu|^2\, |\beta_{\nu'}|^2 \tag{6.23}$$

[12] The summation convention is dropped in the following.

if we restrict ourselves to moderate field strengths. The constants a_ν and

$$b_{\nu\nu'} = b_{\nu'\nu} \tag{6.24}$$

are real constants. In order to meet the natural boundary conditions for W_1^s, at least one of the inequalities

$$b_{\nu\nu'} > 0\,; \qquad b_{\nu\nu}\, b_{\nu\nu'} > b_{\nu\nu'}^2 \tag{6.25}$$

must hold. The coefficients a_ν change sign, if the intensity of the external pump, described by the parameter λ, passes the threshold value λ_ν. λ_ν gives the threshold of mode ν in the absence of all other modes. Therefore, we may put

$$a_\nu = \alpha_\nu(\lambda - \lambda_\nu) \tag{6.26}$$

with positive constants α_ν. An expression of the form (6.23) has also been obtained from the microscopic theory [57], and was used in [56] to investigate the dynamics of a two-mode laser. The shape of the potential (6.23) may take on quite different forms depending on the relative size of the various coefficients. Some typical cases for two modes are shown in Figs. 11–14. Well below the threshold of the first mode $W_1^s \sim \exp(-\phi^s)$ is a multi-dimensional Gaussian, centered around $\beta_\nu = 0$ (cf. Fig. 11). Passing through the threshold of the first mode, the Gaussian becomes first very broad and finally the term $b_{11}|\beta_1|^4$ in Eq. (6.23) has to be taken into account in order to determine the form of W_1^s (cf. Fig. 12). If the pumping is further increased, the second mode could pass the threshold, if it were not suppressed by the presence of the first mode (cf. Fig. 13). For sufficiently hard pumping the second mode will finally start oscillating at a threshold which is determined by the bare threshold of the second mode and the intensity of the first mode (Fig. 14). The next modes will show a similar behaviour.

The form of the potential (6.23) can be tested experimentally by photocount experiments in which photons coming from different modes are counted separately.

In order to derive equations of motion from the potential (6.23) we assume that detailed balance holds in the stationary state. The physical meaning of this assumption will be considered later. The diffusion matrix is assumed to be diagonal with respect to different modes. Furthermore, it is assumed to be constant and to preserve the phase angle invariance of the modes. Applying Eq. (4.23) to the potential (6.23), we obtain

$$\dot{\beta}_\nu = J_\nu - q_\nu\, \partial\phi^s / \partial\beta_\nu^* + F_\nu(t) \tag{6.27}$$

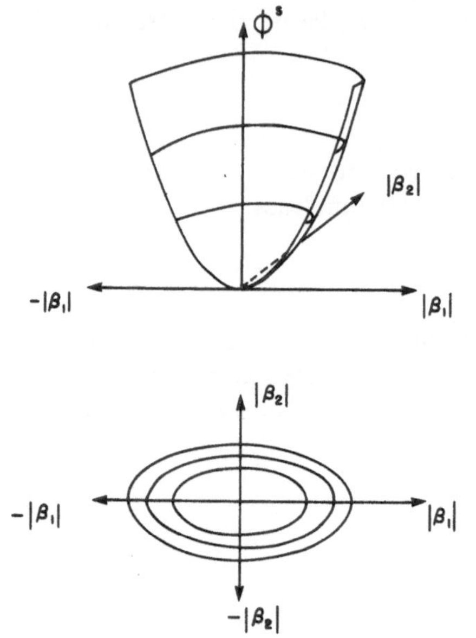

Fig. 11. The potential (6.23) for two modes below threshold ($a_1, a_2 < 0$; $a_1 = \frac{1}{4}a_2$; schematic plot)

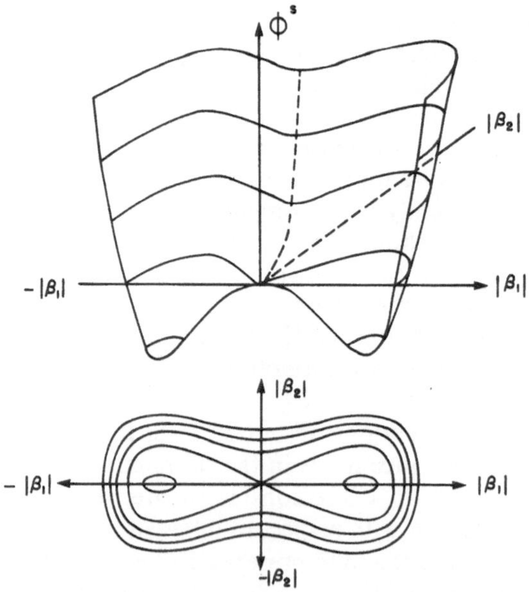

Fig. 12. The potential (6.23) for two modes. One mode (β_1) is above threshold, the other mode is below threshold ($a_1 > 0$; $a_2 < 0$; $a_1 = -\frac{1}{4}a_2$; schematic plot)

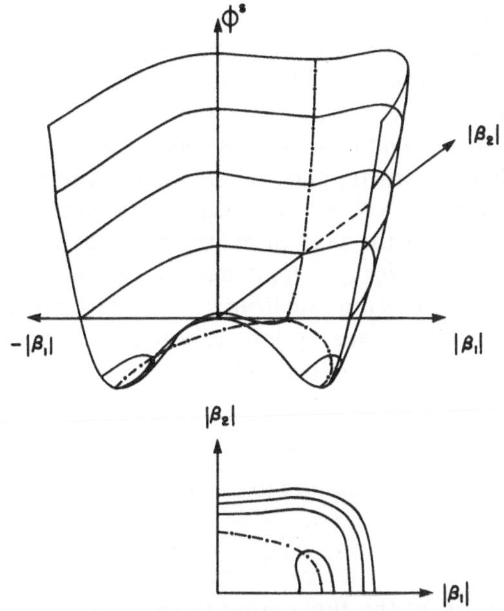

Fig. 13. The potential (6.23) for two modes. One mode (β_1) is above, the other mode is below threshold. The second mode is suppressed by the first mode (a_1, $a_2 > 0$; schematic plot)

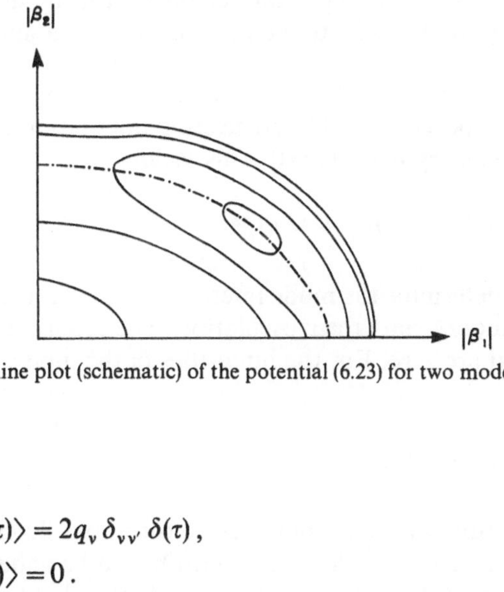

Fig. 14. Contour line plot (schematic) of the potential (6.23) for two modes above threshold

with

$$\langle F_\nu \rangle = 0 \,,$$

$$\langle F_\nu^*(t)\, F_{\nu'}(t+\tau) \rangle = 2q_\nu\, \delta_{\nu\nu'}\, \delta(\tau) \,, \tag{6.28}$$

$$\langle F_\nu(t)\, F_{\nu'}(t+\tau) \rangle = 0 \,.$$

The reversible drift J_ν may be written as a power series in the amplitudes

$$J_\nu = -i\Delta\omega_\nu\, \beta_\nu - i \sum_{\nu'} \Omega_{\nu\nu'}\, |\beta_{\nu'}|^2\, \beta_\nu \tag{6.29}$$

$\Omega_{\nu\nu'}$ and $\Delta\omega_\nu$ are found to be real, if the time reversal symmetry and the phase symmetry is used. $\Delta\omega_\nu$ and $\Omega_{\nu\nu'}$ describe frequency shifts due to the linear response and to nonlinear saturation effects. Eq. (6.24) represents a power series expansion of $\dot{\beta}_\nu$ in terms of the mode amplitudes up to the third order. This expansion is general, apart from the restriction to complete phase symmetry and apart from the symmetry relation (6.24). Thus, Eq. (6.24) is the necessary and sufficient condition for the validity of detailed balance in the present example. From Eq. (6.27) we realize that the quantity $-q_\nu b_{\nu\nu'}|\beta_\nu|^2|\beta_{\nu'}|^2$ has the physical meaning of an induced normalized transition rate from ν' to mode ν. Since the coefficients for induced emission and absorption are equal, the normalized transition rate in the opposite direction has to be equal

$$q_\nu b_{\nu\nu'} = q_{\nu'} b_{\nu'\nu} . \tag{6.30}$$

Therefore, the symmetry relation (6.24) holds, if

$$q_\nu = q_{\nu'} , \tag{6.31}$$

i.e., if the intensity of the fluctuating forces is equal for all modes. In general, this intensity is due to spontaneous emission into the modes. Eq. (6.31) is satisfied, if the spontaneous emission is, at least approximately, constant over the spectral region of the modes. An explicit treatment of two-mode oscillation under the same assumptions was given in [56].

We now consider in somewhat more detail the states which are described by the potential (6.23). Neglecting fluctuations, the stationary states are obtained by $\partial\phi^s/\partial\beta_\nu = 0$. This yields

$$\lambda|\beta_\nu|^2 = \lambda_\nu|\beta_\nu|^2 + 2\alpha_\nu^{-1}\sum_{\nu'} b_{\nu\nu'}|\beta_{\nu'}|^2|\beta_\nu|^2 \tag{6.32}$$

from which we determine the mode intensities, and the threshold values λ at which new modes start their oscillation. The intensities of all modes below threshold are zero. For the intensities of the modes above threshold we obtain

$$|\beta_\nu|^2 = \tfrac{1}{2}\sum_{\nu'} b_{\nu\nu'}^{-1}\alpha_{\nu'}(\lambda - \lambda_{\nu'}) \tag{6.33}$$

where the sum runs over all modes above threshold. Each time when a new mode passes its threshold, a new term has to be added on the right hand side of Eq. (6.33). As a result, at each threshold, the intensities $|\beta_\nu|^2$ have a discontinuous derivative with respect to λ. The pump intensity, which is required to carry mode ν through its threshold if the modes $1 \ldots (\nu-1)$ are already above threshold, is obtained by putting $|\beta_\nu|^2 \geqq 0$.

$$\lambda \geqq \frac{\sum\limits_{\nu'=1}^{\nu} b_{\nu\nu'}^{-1} \alpha_{\nu'} \lambda_{\nu'}}{\sum\limits_{\nu'=1}^{\nu} b_{\nu\nu'}^{-1} \alpha_{\nu'}} . \tag{6.34}$$

This relation can also be used to determine the number of oscillating modes ν, if λ is given. The results of Section 3 may be used to decide whether the stationary state, described by Eq. (6.32), is stable. Since the reversible drift (6.29) and the potential (6.23) satisfy Eq. (3.9), $\phi^s - \phi^s_{\min}$ is a Lyapunoff function of Eq. (6.27) (the fluctuations F_ν are still neglected). The trajectories in the stationary state obey the equations

$$\dot{\beta}^s_\nu = J_\nu(\{\beta^s\}, \{\beta^{*s}\}) \tag{6.35}$$

and are stable against all deviations from these equations since these deviations increase ϕ^s. This result is very easily obtained here and agrees with the less general and more complicated linear stability analysis of the microscopic theory [58]. We now also take into account the fluctuations described by Eq. (6.23), by analyzing the threshold behaviour of mode ν under the condition, that the modes $1 \ldots \nu-1$ are above threshold already. In the vicinity of its threshold, mode ν will have fluctuations with much longer life time than the fluctuations of the other modes. Therefore, we replace the intensities of all other modes by constant parameters $I_{\nu'}$ and obtain

$$\phi^s = -\left(a_\nu - 2 \sum_{\nu'=1}^{\nu-1} b_{\nu\nu'} I_{\nu'}\right) |\beta_\nu|^2 + b_{\nu\nu} |\beta_\nu|^4 . \tag{6.36}$$

This potential has the same form as the potential of a single mode laser. The presence of the $\nu-1$ other modes manifests itself only in the shift of the threshold value, as discussed in Eq. (6.34). Hence, each single instability leading to a new mode is very similar to the single mode case. This result is quite general and depends only on the condition that the thresholds of different modes are well separated from each other.

We close this section by pointing out an interesting analogy between the present example of multimode laser action and turbulence in hydrodynamics. The onset of turbulence has been analyzed by Landau [59] as a succession of instabilities of different modes of the velocity field with independent phases. Each instability brings in a new randomly phased mode of higher frequency and smaller wavenumber and increases the number of arbitrary phases by 1. The resulting motion is highly irregular and is quasi-periodic. In our example we also have a succession of instabilities, each introducing a new arbitrary phase. The total electric field is given by the expansion

$$E(x,t)=\sum_{\nu}(\beta_\nu(t)\exp(-i\omega_\nu t)f_\nu(x)+\text{c.c.}) \tag{6.37}$$

where $f_\nu(x)$ are the resonator modes and ω_ν are the resonator frequencies. Each term in Eq. (6.37) contains an arbitrary phase. The total field $E(x,t)$ is quasi-periodic and consists of a statistical sequence of fluctuation pulses [60].

6.3. Multimode Laser with Mode-Locking

In many cases, different laser modes are coupled, not only by their intensities, but also by their phases. This coupling generally occurs when satellites of laser modes, which are in resonance with neighbouring modes, are created by external or internal modulation [8]. Due to the phase coupling the different modes interfere and produce periodic pulse trains. If the frequency difference between the phase-coupled modes is small, one may obtain a "frequency locking", i.e., a composite oscillation of the mode and its satellite with equal frequency. Typical examples of frequency locking occur in lasers with Zeemann splitted transitions [61], or in lasers with a coupling between the axial and the closely neighbouring nonaxial modes (e.g. due to spatial inhomogeneities, cf. [62]). We start by making the same general assumptions as in the beginning of 6.2. In particular we assume that the dynamics is completely described by the mode amplitudes. Furthermore, we restrict ourselves to the case of moderate amplitudes, so that we may expand ϕ^s in powers of the mode amplitudes. Averaging over times which are long compared to an optical period we obtain

$$\phi^s=-\sum_{\nu,\nu_0}a_{\nu\nu_0}\beta_\nu^*\beta_{\nu_0}+\sum_{\nu_1\nu_2\nu_3\nu_4}b_{\nu_1\nu_2\nu_3\nu_4}\beta_{\nu_1}^*\beta_{\nu_2}^*\beta_{\nu_3}\beta_{\nu_4} \tag{6.38}$$

where higher order terms were neglected. We note that the products occurring in Eq. (6.38) have to be time-independent in order to survive the time average. Therefore, we have resonance between the interacting modes with frequencies $\Delta\omega_\nu$

$$\Delta\omega_{\nu_1}+\Delta\omega_{\nu_2}=\Delta\omega_{\nu_3}+\Delta\omega_{\nu_4}\,. \tag{6.39}$$

Furthermore, the frequencies occurring in the first term of Eq. (6.38) have to coincide with the frequencies of the external forces. The potential (6.38) has a number of phase symmetries, since the phases of the modes have to fulfill the relations

$$\begin{aligned} &\varphi_\nu + \varphi_{\nu_0} = 2n\pi \\ &\varphi_{\nu_1} + \varphi_{\nu_2} + \varphi_{\nu_3} + \varphi_{\nu_4} = 2m\pi \end{aligned} \qquad (n, m \text{ integers}) \tag{6.40}$$

and are arbitrary otherwise. The coefficients $a_{\nu\nu_0}$ and $b_{\nu_1\nu_2\nu_3\nu_4}$ in Eq. (6.38) have to fulfill the symmetry relations

$$a_{\nu\nu_0} = a^*_{\nu_0\nu}, \tag{6.41}$$

$$\begin{aligned} b_{\nu_1\nu_2\nu_3\nu_4} &= b_{\nu_1\nu_2\nu_4\nu_3} \\ b_{\nu_1\nu_2\nu_3\nu_4} &= b^*_{\nu_3\nu_4\nu_1\nu_2}. \end{aligned} \tag{6.42}$$

In the following we specialize Eq. (6.38) for different cases and make contact with the microscopic theory.

a) Self-Locking of Phases

In this case, no external force acts on the system, apart from the usual pump. The coupling between the phases of different axial laser modes is, in this case, due to the nonlinear mode interaction and must be contained in the 4th order terms of Eq. (6.38). Therefore we obtain

$$\phi^s = -\sum_\nu a_\nu |\beta_\nu|^2 + \sum_{\nu_1\nu_2\nu_3\nu_4} b_{\nu_1\nu_2\nu_3\nu_4} \beta^*_{\nu_1}\beta^*_{\nu_2}\beta_{\nu_3}\beta_{\nu_4} \tag{6.43}$$

Assuming detailed balance we may derive equations of motion by applying Eq. (4.23). We get

$$\dot{\beta}_\nu = J_\nu - q_\nu \, \partial\phi^s/\partial\beta^*_\nu + F_\nu. \tag{6.44}$$

J_ν is obtained from the power series

$$J_\nu = -i\Delta\omega_\nu \beta_\nu - i \sum_{\nu_2\nu_3\nu_4} C_{\nu\nu_2\nu_3\nu_4} \beta^*_{\nu_2}\beta_{\nu_3}\beta_{\nu_4} \tag{6.45}$$

where, again, an average over times long compared to the optical period has been taken. The parameters $\Delta\omega_\nu$ and $C\ldots$ have to be real in order to give J_ν the correct time reversal transformation behaviour. Furthermore, Eq. (6.45) implies the symmetry

$$C_{\nu\nu_2\nu_3\nu_4} = C_{\nu\nu_2\nu_4\nu_3}. \tag{6.46}$$

From Eq. (4.15) we obtain

$$a_\nu C_{\nu\nu_2\nu_3\nu_4} = a_{\nu_4} C_{\nu_4\nu_3\nu_2\nu}. \tag{6.47}$$

Eqs. (6.42), (6.46), (6.47) are the conditions of detailed balance in the present case. These symmetry relations are much more restrictive in the present case than they were in the case of intensity coupling. The comparison with the microscopic theory [57] shows that the symmetry relations are approximately fulfilled, if all modes lie sufficiently close to the center of the homogeneously or inhomogeneously broadened line, and if exact resonance between these modes exists. A considerable simplification of the foregoing analysis is possible in all cases, in which the amplitudes r_ν of

$$\beta_\nu = r_\nu \exp - i\varphi_\nu \tag{6.48}$$

can be considered as stabilized constants and only the motion of the phase φ_ν has to be considered. In this case, Eq. (6.43) reduces to

$$\phi^s = \sum_{\nu_1\nu_2\nu_3\nu_4} B_{\nu_1\nu_2\nu_3\nu_4} \cos(\varphi_{\nu_1} + \varphi_{\nu_2} - \varphi_{\nu_3} - \varphi_{\nu_4} - \psi_{\nu_1\nu_2\nu_3\nu_4}) \tag{6.49}$$

where

$$B_{\nu_1\nu_2\nu_3\nu_4} = 2|b_{\nu_1\nu_2\nu_3\nu_4}|\, r_{\nu_1} r_{\nu_2} r_{\nu_3} r_{\nu_4} \tag{6.50}$$

and

$$\exp - i\psi_{\nu_1\nu_2\nu_3\nu_4} = b_{\nu_1\nu_2\nu_3\nu_4}/|b_{\nu_1\nu_2\nu_3\nu_4}|\,. \tag{6.51}$$

The phases which are realized with maximum probability are obtained from the extremum principle

$$\delta\phi^s = 0\,, \qquad \delta^2\phi^s > 0\,. \tag{6.52}$$

An extremum principle of maximum gain has been introduced previously by intuition, in order to study phase locked lasers [63]. Our extremum principle (6.52), whenever it is applicable, is equivalent to this principle of maximum gain. As a specific example we consider the case of 3 interacting modes, which are tuned to satisfy Eq. (6.39)

$$\delta \equiv 2\Delta\omega_2 - \Delta\omega_1 - \Delta\omega_3 = 0\,. \tag{6.53}$$

From Eq. (6.49) we obtain

$$\phi^s = A\cos(\Delta\psi - \psi_0) \tag{6.54}$$

where

$$\Delta\psi = 2\varphi_2 - \varphi_3 - \varphi_1 + \delta t\,. \tag{6.55}$$

The distribution, given by Eq. (6.54) has also been obtained from the microscopic theory [57, 64]. The equation of motion (4.23) derived from the potential (6.54) has the form

$$\Delta\dot\psi = \delta + \tfrac{1}{2} Aq \sin(\Delta\psi - \psi_0) + F(t) \tag{6.56}$$

with

$$\langle F(t)\rangle = 0\,; \qquad \langle F(t)\,F(t')\rangle = q\delta(t-t')\,. \tag{6.57}$$

The Langevin equation (6.56) has been analyzed previously in all detail [64]. We use this analysis to consider the consequences of a finite detuning δ, which violates Eq. (6.39). The Fokker-Planck equation, corresponding to Eqs. (6.56), (6.57) has the stationary solution

$$W_1^s(\Delta\psi) = N_0 \exp[2\delta\Delta\psi/q - A\cos(\Delta\psi - \psi_0)] \int_{\Delta\psi}^{\Delta\psi+2\pi} \cdot \exp[-2\delta\varphi/q + A\cos(\varphi-\psi_0)]\, d\varphi \tag{6.58}$$

where N_0 is a normalization constant. For $\delta = 0$ this solution is reduced to Eq. (6.54). Introducing this solution into Eq. (2.22), we obtain

$$r^s(\Delta\psi) = q[2N_0 W_1^s(\Delta\psi)]^{-1}(1-\exp(-4\pi\,\delta/q))\,. \tag{6.59}$$

Eq. (6.59) shows, that $\delta \neq 0$ will induce a nonvanishing drift velocity in the stationary state. Hence, $\delta = 0$ is the condition for detailed balance in the present case.

b) *Forced Locking of Phases*

Mode locking can be forced by an external modulation of the losses or the gain. If the modulation frequency coincides with the difference in frequency of neighbouring axial modes, the gain of the generated satellite modes will depend on the phases of these neighbouring modes. Hence a phase coupling of modes with initially uncorrelated phases is produced. In this case, the most important phase coupling is already contained in the bilinear terms of Eq. (6.38). The phase coupling in the higher order terms is then of minor importance and is left aside here. ϕ^s has the form

$$\phi^s = \sum_\nu \left[-a_\nu|\beta_\nu|^2 - a_\nu^{(1)}(\beta_{\nu+1}^*\beta_\nu + \beta_\nu^*\beta_{\nu+1}) + \sum_{\nu'} b_{\nu\nu'}|\beta_\nu|^2\,\beta_{\nu'}|^2\right] \tag{6.60}$$

where $a_\nu^{(1)}$ is real and is proportional to the external locking force. In order to see how $a_\nu^{(1)}$ leads to a locking of phases in the presence of many modes, we consider the index ν as a continuous variable and obtain

$$\phi^s = \int d\nu[-a(\nu)|\beta(\nu)|^2 + a^{(1)}(\nu)\,|\partial\beta(\nu)/\partial\nu|^2 + \int d\nu'\, b(\nu,\nu')|\beta(\nu)\,\beta(\nu')|^2] \tag{6.61}$$

where we defined

$$\begin{aligned} a^{(1)}(\nu) &= a_\nu^{(1)}\varrho_\nu \\ a(\nu) &= (a_\nu - a_\nu^{(1)} - a_{\nu-1}^{(1)})\,\varrho_\nu \end{aligned} \tag{6.62}$$

and $\varrho_\nu d\nu$ is the number of modes in the interval $(\nu, \nu + d\nu)$. Eq. (6.61) shows that, for $a^{(1)}(\nu) > 0$, ϕ^s becomes smaller if $|\partial\beta(\nu)/\partial\nu|$ is decreased. Hence, the term containing $a^{(1)}$ tends to make the amplitude $\beta(\nu)$ uniform with respect to the index ν. Since the electric field is given by the fourier transform with respect to ν, we obtain a pulse sharpening in the time domain. The number of modes whose phases are locked by $a^{(1)}(\nu)$ can be estimated as follows: Assuming constant amplitudes r_ν of $\beta(\nu)=r(\nu)\exp-i\varphi(\nu)$, the probability density W_1^s is given by

$$W_1^s \sim \exp\left[-\int d\nu\, a^{(1)}(\nu)\, r^2(\nu)\,(\partial\varphi(\nu)/\partial\nu)^2\right]. \tag{6.63}$$

This probability density functional of the stochastic phase $\varphi(\nu)$ describes a Brownian motion of the phase along a coordinate ν. The quantity

$$\Delta\nu = a^{(1)}(\nu)\, r^2(\nu) \tag{6.64}$$

defines a coherence interval, since for $|\nu_1-\nu_2| \ll \Delta\nu$ the phases of two given modes ν_1, ν_2 are coherent, whereas for $|\nu_1-\nu_2| \gg \Delta\nu$ they are completely at random. The size of the coherence interval is proportional to the modulation strength and to the mode intensity. The equations of motion derived from (6.60) by applying (4.23) are

$$\dot\beta_\nu = q_\nu\left(a_\nu\beta_\nu + a_{\nu-1}^{(1)}\beta_{\nu-1} + a_\nu^{(1)}\beta_{\nu+1} - 2\sum_{\nu_1} b_{\nu\nu_1}|\beta_{\nu_1}|^2\beta_\nu\right) + F_\nu \tag{6.65}$$

where we assumed a constant and diagonal diffusion matrix, and, for simplicity, disregarded frequency shifts. The latter could be taken into account if necessary. Eq. (6.65) can be compared with equations of a microscopic theory given in [65]. Agreement is obtained if we put

$$q_\nu a_\nu = -\kappa + g_\nu\,; \qquad q_\nu a_\nu^{(1)} = \kappa_c\,; \qquad 2q_\nu b_{\nu\nu'} = g_\nu g_{\nu'} \tag{6.66}$$

κ is the loss which, in [65], is assumed to be equal for all modes; g_ν is the gain of mode ν; κ_c is the amplitude of the loss modulation. This special choice of coefficients[13] allows us to put Eq. (6.65) for $\dot\beta_\nu=0$ into the form of a linear eigenvalue equation

$$\kappa B_\nu^M - \kappa_c(B_{\nu+1}^M + B_{\nu-1}^M) = g_\nu G_0(M)\, B_\nu^M \tag{6.67}$$

for the eigenfunction B_ν^M and the real eigenvalues $G_0(M)$. β_ν is given as a linear superposition

$$\beta_\nu = \sum_M C_M B_\nu^M\,. \tag{6.68}$$

The apparent nonlinearity of Eq. (6.65) is hidden in the linear Eq. (6.67), because, in addition, $G_0(M)$ has to satisfy

$$G_0(M) = 1 - 2\sum_\nu g_\nu |B_\nu^M|^2\,. \tag{6.69}$$

[13] The following results were obtained previously by H. Geffers, University of Stuttgart, in unpublished calculations based on a Fokker-Planck equation.

The eigenfunctions B_ν^M satisfy an orthogonality relation with the weight function g_ν

$$\sum_\nu g_\nu B_\nu^M B_\nu^{*M'} = \tfrac{1}{2}(1-G_0(M))\,\delta_{MM'}\,. \tag{6.70}$$

The eigenvalues $G_0(M)$ are real. Expressing ϕ^s in terms of the new variables C_M we obtain

$$\phi^s = -\sum_M \tfrac{1}{2}|C_M|^2(1-G_0(M))^2 + \Big(\sum_M \tfrac{1}{2}(1-G_0(M))\,|C_M|^2\Big)^2. \tag{6.71}$$

This gives us the distribution of the coefficients C_M in the stationary state. The most probable configuration is given by

$$\partial\phi^s/\partial C_M = 0 \tag{6.72}$$

which has the solution

$$|C_M|^2 = \delta_{MM_0}\,. \tag{6.73}$$

Therefore, most likely the configuration $B_\nu^{M_0}$ is excited in the stationary state. In order to determine the most probable value of M_0, we introduce

$$C_M = \delta_{MM_0} + \delta C_M \tag{6.74}$$

into Eq. (6.71), obtaining

$$\phi^s = -\tfrac{1}{4}(1-G_0(M))^2 + \tfrac{1}{2}\sum_M (1-G_0(M))\,(G_0(M)-G_0(M_0))\,|\delta C_M|^2\,. \tag{6.75}$$

From Eq. (6.70) we know that $1-G_0(M) \geqq 0$. Hence the last term in Eq. (6.75) is positive and the solution (6.73) is stable only if M_0 minimizes $G_0(M)$. This configuration gives the absolute minimum of ϕ^s. From our general theory we know that this absolute minimum is stable, whereas all other configurations are unstable. The phase of C_{M_0} is not determined by this argument, and a diffusion of the phase of C_{M_0} will take place due to fluctuations. Eq. (6.75) gives an expression for the probability density of the excitation of other configurations $B_\nu^M (M \neq M_0)$ in the stationary state. The configurations with smallest and with largest $\sum_\nu g_\nu |B_\nu^M|^2$ are excited most likely. In view of Eq. (6.68) only the latter give an important contribution to the total field.

c) Frequency Locking

Frequency locking indicates the oscillation of different modes with equal frequency. These modes would have slightly different frequencies in the noninteracting case. In typical cases the nearly degenerate modes arise from Zeemann splitting in a weak magnetic field [61] or from the excitation of closely spaced nonaxial modes [62]. As a result of frequency locking, the mode structure of the filled resonator differs from the mode structure of the unfilled resonator. By application of our theory it is possible, to give an expression for the probability density of finding a certain amplitude of a mode of the unfilled resonator. As in the case of phase locking discussed in 6.3a, b, we may distinguish two cases: locking due to nonlinear, and to linear mode coupling. The analysis of both cases completely parallels the analysis given in 6.3a for the nonlinear locking and in 6.3b for the forced linear locking. However, in the frequency locking case we no longer distinguish different locked modes by their different frequencies. We rather have to use different mode characteristics like the polarization (in the case of the Zeemann splitted laser) or the spatial mode structure (in the case of nonaxial modes). Since the analysis is similar to the considerations of Section 6.3a, b, we discuss here only the simple example of the linearly induced frequency locking of two modes, which has been discussed in the literature in the frame of the microscopic theory [66]. We obtain from Eq. (6.38), by neglecting nonlinear phase coupling,

$$\phi^s = -\sum_\nu a_\nu |\beta_\nu|^2 - a^{(1)} \beta_1^* \beta_2 - a^{(1)*} \beta_1 \beta_2^* + \sum_{\nu\nu'} b_{\nu\nu'} |\beta_\nu|^2 |\beta_{\nu'}|^2 \tag{6.76}$$

which gives the probability density, $W_1^s \sim \exp(-\phi^s)$, of finding a certain amplitude of the modes of the empty resonator. Eq. (6.76) reproduces an end result of the microscopic theory [66] which, in the present case, turns out to be very involved. For well stabilized amplitudes r_ν of $\beta_\nu = r_\nu \exp - i\varphi_\nu$, Eq. (6.76) reduces to

$$W_1^s \sim \exp(-2|a^{(1)}| r_1 r_2 \cos(\Delta\psi - \psi_0)) \tag{6.77}$$

with

$$\Delta\psi = \varphi_1 - \varphi_2 \tag{6.78}$$

and

$$\exp i\psi_0 = a^{(1)}/|a^{(1)}| . \tag{6.79}$$

Some further implications of Eq. (6.76) and the possibilities of putting this result to experimental test have been discussed in [66].

6.4. Light Propagation in an Infinite Laser Medium

The statistical analysis of the complete space-time behaviour of laser fields became important after the discovery of ultrashort light pulses. The description of these pulses in the mode picture is no longer economical, since too many modes participate in the nonlinear interaction. The general analysis of the statistics of ultrashort light pulses is one of the most important unsolved problems in quantum optics. This problem can not immediately be attacked with the methods described here, since it involves many rapidly varying variables, invalidating the Markoff assumption. However, if the fluctuation problem can be reformulated in terms of a small number of slowly varying space-time dependent fields, one may apply our theory. Two examples are given below. First we generalize the considerations of section 6.1 from single mode operation to the propagation of a space-time dependent field in a one-dimensional laser medium of infinite length. This generalization is significant, since it shows how close the analogy between the laser threshold and systems near a critical point of phase transitions really is. A microscopic theory of this example was given in [45].

In the vicinity of the laser threshold we describe the system by the slowly varying complex amplitude $\beta(x, t)$, which now also depends on the space variable x. We determine ϕ^s by an expansion in powers of this amplitude, observing the general rules given in 3.2. We now have to take into account the space variation of the amplitude as well. Assuming slow spatial variation, we retain only the lowest order term in $\partial\beta/\partial x$ and obtain

$$\phi^s = \int \mathrm{d}x[-a|\beta(x)|^2 + b|\beta(x)|^4 + \mathrm{d}|\partial\beta(x)/\partial x|^2] \tag{6.80}$$

where

$$b>0, \quad d>0, \tag{6.81}$$

$$a=\alpha(\lambda-\lambda_c). \tag{6.82}$$

Expression (6.80) is the well known basis of the Landau theory of phase transitions with a space dependent order parameter [67]. The relative magnitude of α and d can be determined if we put $\lambda=0$ and neglect the fourth order term in Eq. (6.80). Then we can calculate the average $\langle\beta^*(x)\,\beta(0)\rangle$ from the distribution $W_1^s \sim \exp(-\phi^s)$ by functional integration. We obtain

$$\langle\beta^*(x)\,\beta(0)\rangle = \langle|\beta|^2\rangle \exp -|x|/\xi_0\,. \tag{6.83}$$

The coherence length

$$\xi_0 = \sqrt{\mathrm{d}/\alpha\lambda_c} \tag{6.84}$$

is the length of the wave packets of spontaneous emission. If the laser atoms have a natural atomic line width $\gamma_\perp$ and if the additional damping in the medium is κ, we have

$$\xi_0 = c(\kappa+\gamma_\perp)^{-1} \tag{6.85}$$

where c is the velocity of light, and

$$d = c^2 \alpha \lambda_c (\kappa + \gamma_\perp)^{-2} \tag{6.86}$$

ξ_0 is larger than 1 cm in optical systems, and hence much longer than the corresponding coherence lengths in superconductors or superfluids. This large coherence length accounts for the fact that spatial fluctuations are of little importance in lasers of typical dimensions. At the same time it reveals one of the reasons for the accuracy of the Landau theory in optical examples. As is well known, the Landau theory becomes exact if the coherence length ξ_0 becomes very large. The potential (6.80) makes it possible to calculate single time expectation values of the field. To this end we define the quantity

$$Q(\beta, \beta^* | \beta', \beta'^*; x - x') \equiv \langle \delta^2(\beta(x) - \beta)\, \delta^2(\beta(x') - \beta') \rangle \tag{6.87}$$

with

$$\delta^2(\beta(x) - \beta) \equiv \delta(\mathrm{Re}\,\beta(x) - \mathrm{Re}\,\beta)\, \delta(\mathrm{Im}\,\beta(x) - \mathrm{Im}\,\beta). \tag{6.88}$$

The average on the right hand side of Eq. (6.87) defines a functional integral of the Wiener type. Instead of doing this integral one can evaluate the "Schrödinger equation"

$$\partial Q/\partial x = d^{-1}\, \partial^2 Q/\partial \beta \partial \beta + (-a|\beta|^2 + b|\beta|^4)\, Q \tag{6.89}$$

which is equivalent to Eq. (6.87) [22, 68]. The time independent form of Eq. (6.89) describes energy eigenstates of a quantum particle with mass $2d$ in the potential, shown in Figs. 4,5. Eq. (6.89) can be solved by approximation procedures or numerical methods familiar from quantum theory. Once Q is determined from Eq. (6.89), one can calculate averages of the form

$$A_{12}(x, x') \equiv \langle F_1(\beta(x), \beta^*(x))\, F_2(\beta(x'), \beta^*(x')) \rangle \tag{6.90}$$

by the relation

$$A_{12}(x, x') = \int d^2\beta\, d^2\beta'\, F_1(\beta, \beta^*)\, F_2(\beta', \beta'^*)\, Q(\beta, \beta^* | \beta', \beta'^*; x - x') \tag{6.91}$$

We don't evaluate the results in this generality here. The most important effect which determines the coherence length of the amplitude $\beta(x, t)$, is the spatial diffusion of the phase of the light, which can be evaluated without solving Eq. (6.89). The same phase diffusion is well known in the theory of 1-dimensional superconductors (cf. e.g. [69]). If we decompose

$$\beta(x) = r \exp - i\varphi(x) \tag{6.92}$$

and assume that r is a space independent constant (which is valid well above threshold), then Eq. (6.80) reduces to

$$\phi^s = d \cdot r^2 \int dx (\partial\varphi/\partial x)^2 . \tag{6.93}$$

From Eq. (6.93) we obtain for the average

$$\langle(\varphi(x) - \varphi(x'))^2\rangle = |x - x'|/\xi \tag{6.94}$$

with the new coherence length

$$\xi = 2dr^2 = \alpha(\lambda - \lambda_c)\, d/b . \tag{6.95}$$

Eq. (6.94) clearly shows that the phase undergoes a spatial diffusion. If the space integral in Eq. (6.93) would be taken over a three dimensional volume, the phase diffusion would vanish and would be replaced by a zero frequency oscillation around a constant value (Goldstone mode).

In the second example of this section we look at the propagation of periodic pulse trains in an infinite laser medium. The spontaneous occurrence of trains of periodic pulses in a medium with translational invariance is again connected with a symmetry changing instability. In fact, this instability, considered in the mode picture, has already been considered in Section 6.3a. Here, we are interested in the state, in which many laser modes are firmly locked to form a periodic train of very short and intense pulses. This state has recently been analyzed in a theory which neglects fluctuations, and stationary periodic pulse trains have been found [70]. We apply our phenomenological theory in order to see how this result is modified, if fluctuations (due to spontaneous emission) are taken into account. We assume, that the intensity of the field can be approximated by the non-fluctuating periodic functions, found in [70], and that phase fluctuations of the field have the most important effect. A fluctuation in the phase of a propagating field is equivalent to a fluctuation in its propagation velocity. If the frame of reference moves with the average propagation velocity of the field, the space dependent phase fluctuations lead to a fluctuating space-time dependent displacement of the field intensity relative to the nonfluctuating state. We may describe this displacement by a "displacement vector" $u(x, t)$ in terms of which the fluctuating field $\beta(x, t)$ is given by

$$\beta(x, t) = \beta_0(x + u(x, t)) . \tag{6.96}$$

$\beta_0(x)$ is the stationary, periodic field when fluctuations are neglected, as given in [70]. The stationary distribution of $u(x)$ can now be found from symmetry arguments. ϕ^s may only depend on spatial derivatives of $u(x)$, since a uniform displacement cannot alter the probability density.

Hence, to the lowest order, ϕ^s is given by

$$\phi^s = \int \lambda(x)\,(\partial u/\partial x)^2\,\mathrm{d}x \qquad (6.97)$$

with

$$\lambda(x) = \lambda(x+R) \qquad (6.98)$$

$\lambda(x)$ must have the periodicity of the nonfluctuating pulse train. This treatment of phase fluctuations in periodic pulse trains is completely equivalent to the problem of displacement fluctuations in one-dimensional crystals, discussed by Landau [71]. As is well known in the case of one-dimensional crystals, the excitation of phonons leads to the destruction of strict periodicity. The same result holds for phase fluctuations in 1-dimensional pulse trains. The destruction of long range order in the pulse train can be realized by calculating the mean square displacement between two points with $|x-x'| \gg R$. We obtain

$$\langle (u(x)-u(x'))^2 \rangle = |x-x'|/\xi \qquad (6.99)$$

with

$$\xi = (2/R)\int_0^R \lambda(x)\,\mathrm{d}x\,. \qquad (6.100)$$

The result (6.99) grows linearly with the distance, indicating a diffusion process which destroys periodicity over distances larger than ξ. Since the process under consideration is again a phase diffusion, the coherence length ξ in Eq. (6.99) can be estimated by Eq. (6.95), where r^2 has to be replaced by a spatial average of the field intensity. Since in the case of pulse trains the field intensity becomes very small between the pulses, the coherence length can become much smaller than in the case of single mode operation. If the coherence length ξ becomes comparable with the pulse period R, a breaking up of the pulse train into a stochastic sequence of fluctuation pulses must take place. This break up corresponds to a transition from the phase locking Region 6.3 to the random phase Region 6.2.

7. Parametric Oscillation

Besides the instabilities encountered in laser active media, there exists another class of instabilities in nonlinear optics. These instabilities are included in passive optical media by shining in a coherent laser field. Therefore, these are also instabilities in stationary nonequilibrium states far from thermal equilibrium. In this section we consider the simplest examples by applying the general theory of part A. In section 8 we consider also more complicated examples.

7.1. The joint Stationary Distribution for Signal and Idler

We consider the nonlinear optical process in which a light quantum with frequency ω_p and wave vector k_p is transformed into two quanta with the frequencies ω_1, ω_2 and the wavenumbers k_1, k_2 [14]. It is assumed that none of the three frequencies is in resonance with excitations of the medium. The basic scattering process is shown in Fig. 15. In the

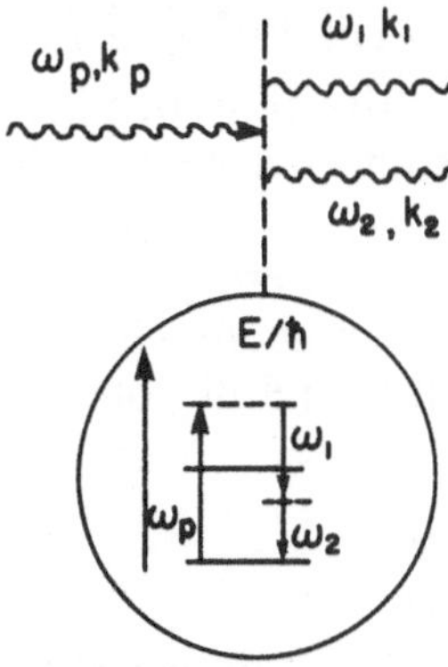

Fig. 15. Second order parametric scattering of light and the corresponding electronic transitions in a two-level atom

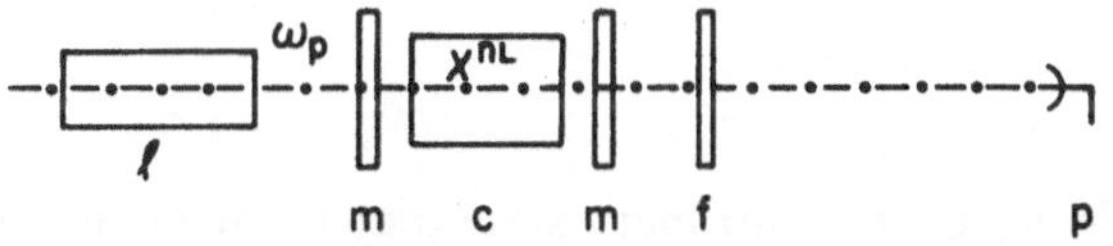

Fig. 16. Scheme of an oscillator, based on induced light scattering in a medium with a field dependent dielectric susceptibility. (*l* laser, *m* mirrors which are transparent at the laser frequency ω_p, but highly reflecting at the oscillator frequencies, *c* crystal with field dependent susceptibility χ^{nl}, *f*-filter absorbing the laser light, *p* photo-detector)

same figure we show the virtual transitions in a two-level system which would give rise to this scattering. A typical experimental set up is shown in Fig. 16; it was first realized by Giordmaine and Miller [81].

We assume perfect frequency matching

$$\omega_1 + \omega_2 = \omega_p \tag{7.1}$$

and phase matching

$$k_1 + k_2 = k_p \tag{7.2}$$

Mirrors which reflect light at the frequencies ω_1 and ω_2, but do not reflect light at the frequency ω_p are employed to reduce the losses at

[14] A general introduction to parametric processes of this kind is given in [72]. See also [9, 11, 37, 44, 73 – 80].

ω_1, ω_2 considerably below the losses at ω_p. Then the complex amplitudes β_1, β_2 of the modes with the frequencies ω_1, ω_2 are the only slowly varying variables of the system. The external parameters $\{\lambda\}$ are represented by the complex amplitude F_p of the pumping laser.

The potential ϕ^s is again determined by a power series expansion with respect to the amplitudes β_1, β_2. Keeping only resonant terms, we obtain in lowest order

$$\phi^s = a_1 |\beta_1|^2 + a_2 |\beta_2|^2 + a_{12} F_p \beta_1^* \beta_2^* + a_{12}^* F_p^* \beta_1 \beta_2 \tag{7.3}$$

where a_1, a_2 are real and positive constants and a_{12} is a complex constant. The terms $\sim F_p, F_p^*$ are time independent, because of the resonance condition (7.1). The constant a_{12} must be proportional to the nonlinear susceptibility giving rise to the scattering process of Fig. 15.

In order to derive equations of motion from Eq. (7.3) we assume that the diffusion matrix is constant and diagonal (cf. Section 6.1) and use Eq. (2.24). The result is

$$\begin{aligned} \dot{\beta}_1 &= -q_1\, \partial\phi^s/\partial\beta_1^* + r_1^s + F_1(t) \\ \dot{\beta}_2 &= -q_2\, \partial\phi^s/\partial\beta_2^* + r_2^s + F_2(t) \end{aligned} \tag{7.4}$$

where q_1 and q_2 are the diagonal elements of the diffusion matrix. The fluctuating forces $F_{1,2}$ have the properties

$$\begin{gathered} \langle F_1 \rangle = \langle F_2 \rangle = 0\,; \quad \langle F_1(t)\, F_2(t') \rangle = \langle F_1(t)\, F_2^*(t') \rangle = 0 \\ \langle F_1^*(t)\, F_1(t+\tau) \rangle = 2q_1\, \delta(\tau) \\ \langle F_2^*(t)\, F_2(t+\tau) \rangle = 2q_2\, \delta(\tau)\,. \end{gathered} \tag{7.5}$$

The drift velocity in the stationary state may be given as a power series expansion in the mode amplitudes. In addition it has to satisfy Eq. (2.20). Up to the accuracy of ϕ^s we obtain

$$\begin{aligned} r_1^s &= -i\alpha_1\, \beta_1 - i\, \delta_1 (a_{12}/a_1)\, F_p\, \beta_2^* \\ r_2^s &= -i\alpha_2\, \beta_2 - i\, \delta_2 (a_{12}/a_2)\, F_p\, \beta_1^* \end{aligned} \tag{7.6}$$

where the coefficients $\alpha_1, \alpha_2, \delta_1, \delta_2$ are real and fulfill the relation

$$\alpha_1 + \alpha_2 = \delta_1 + \delta_2\,. \tag{7.7}$$

Clearly, $\alpha_1, \alpha_2, \delta_1, \delta_2$ describe detuning effects, which cannot be present in our case, since we assumed exact resonance in Eq. (7.1). Therefore, we may put

$$\alpha_1 = \alpha_2 = \delta_1 = \delta_2 = 0 \tag{7.8}$$

in the following. In the stationary state the amplitudes fluctuate around the state

$$\beta_1 = \beta_2 = 0. \tag{7.9}$$

The state (7.9) becomes unstable, if the bilinear form in β_1, β_2, Eq. (7.3), is no longer positive definite. This happens if the secular equation

$$\lambda^2 = (a_1 q_1 + a_2 q_2)\,\lambda + a_1 a_2 q_1 q_2 - |a_{12} F_p|^2\, q_1 q_2 = 0 \tag{7.10}$$

has a negative root

$$\lambda_{1,2} = \tfrac{1}{2}(a_1 q_1 + a_2 q_2) \pm \sqrt{\tfrac{1}{4}(a_1 q_1 - a_2 q_2)^2 + |a_{12} F_p|^2\, q_1 q_2}\,, \tag{7.11}$$

i.e. if

$$|F_p|^2 > a_1 a_2 / |a_{12}|^2\,. \tag{7.12}$$

In order to determine ϕ^s in the vicinity of this instability, we diagonalize Eq. (7.3) by the transformation

$$\begin{aligned} v_1 &= \frac{(q_1 a_1 - \lambda_2)\,\beta_1 + q_1 a_{12} \beta_2^*}{\sqrt{q_1^2 a_1 (a_1 q_1 + a_2 q_2)}} \\ v_2 &= \frac{(q_2 a_2 - \lambda_2)\,\beta_1 - q_1 a_{12} \beta_2^*}{\sqrt{q_1 q_2 a_2 (a_1 q_1 + a_2 q_2)}} \end{aligned} \tag{7.13}$$

and obtain

$$\phi^s = \lambda_1 |v_1|^2 + \lambda_2 |v_2|^2\,. \tag{7.14}$$

In (7.13), (7.14) we used the approximations

$$\lambda_1 = a_1 q_2 + a_2 q_2\,, \qquad \lambda_2 = \frac{q_1 q_2 (a_1 a_2 - |F_p a_{12}|^2)}{a_1 q_1 + a_2 q_2} \tag{7.15}$$

valid in the threshold region. Eq. (7.15) shows that $\lambda_2 \ll \lambda_1$ in the threshold region. The instability occurs only with respect to the mode v_2, whereas v_1 is heavily damped at the threshold. The form of ϕ^s in the vicinity of $v_1 = v_2 = 0$, for $|F_p|^2$ slightly above the threshold (7.12), is shown in Fig. 17. In order to describe the threshold region completely, we have to add higher order terms to the potential ϕ^s. Since only v_2 becomes unstable, we need only add higher order terms with respect to v_2. We obtain

$$\phi^s = \lambda_1 |v_1|^2 + \lambda_2 |v_2|^2 + b |v_2|^4 \tag{7.16}$$

with real, positive b. Eq. (7.16) gives the joint stationary distribution for both parametrically excited modes (signal and idler) in the threshold region. This result has not yet been obtained by the microscopic theory of stationary parametric oscillation [75 – 80]. However, some results of the microscopic theory are contained in Eq. (7.16) as special cases and we now consider them.

i) The distribution for the "signal" amplitude [78]. This distribution is obtained from Eq. (7.16) by integrating over β_2, or, even more simply, by using the fact that v_1 is heavily damped compared to v_2 and can be put equal to zero. This yields

$$\beta_2 = \frac{\lambda_2 - q_1 a_1}{q_1 a_{12}} \beta_1 , \tag{7.17}$$

$$v_2 = \frac{q_1 a_1 + q_2 a_2 - 2\lambda_2}{\sqrt{q_1 q_2 a_2 (a_1 q_1 + a_2 q_2)}} \beta_1 , \tag{7.18}$$

$$\phi^s = \frac{\lambda_2}{q_1 q_2 a_2} (a_1 q_1 + a_2 q_2) |\beta_1|^2 + \frac{b(a_1 q_1 + a_2 q_2)^2}{(q_1 q_2 a_2)^2} |\beta_1|^4 . \tag{7.19}$$

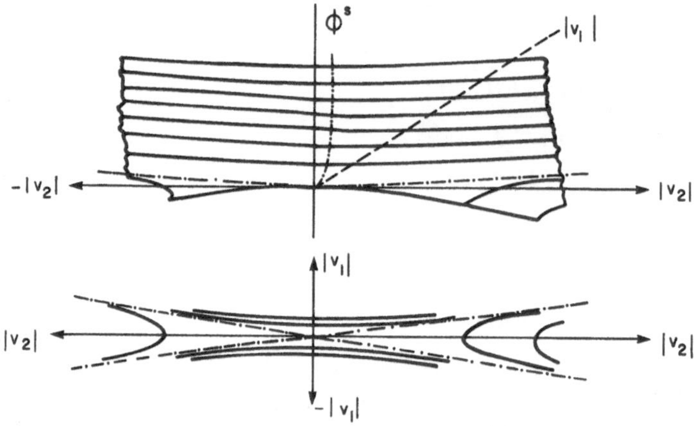

Fig. 17. The potential (7.14) slightly above threshold, in the vicinity of $v_1 = v_2 = 0$. The potential has a sharp minimum with respect to $|v_1|$

The potential (7.19) has the same form as the potential for a single mode laser. This had to be expected in view of the fact that the same basic principles govern both instabilities. The result (7.19) has also been found in the microscopic theory [78].

ii) The joint distribution of signal and idler, in the case of equal damping, has been obtained recently from the microscopic theory (cf. [80] and Section 8). This special case is obtained from Eq. (7.16) by putting

$$a_1 = a_2 = a ; \qquad q_1 = q_2 = q \tag{7.20}$$

which yields

$$\lambda_{1,2} = aq \pm |a_{12} F_p| \, q \tag{7.21}$$

$$v_{1,2} = (2q)^{-1/2} (\beta_1 \pm a_{12} F_p \beta_2^* / |a_{12} F_p|) . \tag{7.22}$$

The result of the microscopic theory [80] reads in our present notation

$$\phi^s = \lambda_1 |v_1|^2 + \lambda_2 |v_2|^2 + b(|v_1|^4 + |v_2|^4 - v_1^2 v_2^{*2} - v_1^{*2} v_2^2)\,. \tag{7.23}$$

The two results (7.16) and (7.23) are the same if

$$\lambda_1 \gg 2b|v_2|^2 \tag{7.24}$$

holds. $|v_2|^2$ can be estimated by taking the maximum of the potential (7.16). Then the condition (7.24) is reduced to $|\lambda_2| \ll \lambda_1$ which, by Eq. (7.21), defines a region around threshold

$$|a - |a_{12}F_p||/2a \ll 1 \tag{7.25}$$

where the phenomenological theory applies.

7.2. Subharmonic Oscillation

Subharmonic oscillation occurs if signal and idler degenerate to one single mode [72, 82], i.e., we have

$$\begin{aligned} \omega_1 &= \omega_2 = \omega_p/2 \\ k_1 &= k_2 = k_p/2\,. \end{aligned} \tag{7.26}$$

This case is contained in the theory of the last section. Since it has some peculiar features of its own, we discuss it separately. In the case of nondegenerate parametric oscillation the phases of β_1 and β_2 are not determined in the stationary state. Only the sum of their phases is locked to the phase of the pump amplitude F_p.

The sum of the phases of signal and idler degenerates to the double of the subharmonic phase which is then locked to the phase of the pump field. Therefore, the double phase of the subharmonic is fixed up to multiples of 2π. The phase itself is then fixed up to multiples of π. Subharmonic generation presents, therefore, an example of a symmetry changing instability, where the symmetry, which is changed, is discrete, rather than continuous, as in our other examples. The minima of ϕ^s in the ordered state will be discontinuously degenerate. Specializing Eq. (7.16) for the present case we find

$$\phi^s = a|\beta|^2 + a_{12}F_p\beta^{*2} + a_{12}^* F_p^* \beta^2 + b|\beta|^4\,. \tag{7.27}$$

The contour lines of the potential (7.27) in the complex β-plane are shown in Fig. 18. In order to obtain the probability density for the absolute value of the amplitude $|\beta|$ alone, we integrate the distribution $W_1 \sim \exp(-\phi^s)$ over the phase of the subharmonic to obtain

$$W_1^s(|\beta|) \sim |\beta|\, I_0(2|a_{12}F_p\beta^2|) \exp(-a|\beta|^2 - b|\beta|^4)\,. \tag{7.28}$$

The normalization has to be carried out by integrating over $|\beta|$ from 0 to ∞. I_0 is the Bessel function with imaginary argument and index 0. If we expand the Bessel function in a power series, keeping only the lowest order, we obtain

$$W_1^s(|\beta|) \sim \exp(-(a-2|a_{12}F_p|)\,|\beta|^2 - b'\,|\beta|^4)\,. \tag{7.29}$$

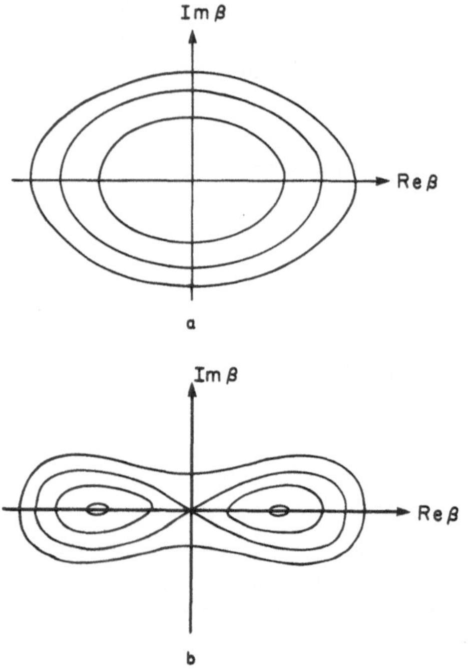

Fig. 18. The potential ϕ^s for subharmonic oscillation, Eq. (7.28) below (a) and above (b) threshold

This result has, again, the same form as the result for the single mode laser. In this case, however, the form (7.29) is not due to a phase angle invariance of the system, which is intrinsic to the system, but rather to complete lack of knowledge of the phase because of integrating over this variable.

8. Simultaneous Application of the Microscopic and the Phenomenological Theory

In the previous sections we gave examples for the application of the general phenomenological theory, set forth in the first part of this paper. The microscopic theory was only used to compare the results, wherever it was possible. In this last chapter we will treat nonlinear optical phenomena by combining the microscopic and the phenomenological approach. The microscopic theory, whose concepts were given in Section 5.2, is used to derive a Fokker-Planck equation for the quasi-probability density of the interesting mode amplitudes. We make use of the phenomenological theory, when we chose a system which has the property of detailed balance with respect to the mode amplitudes. This choice allows us to employ the potential conditions of Section 4.2 to write down the stationary quasi-probability density for the considered process. In section 8.1 we introduce the general scattering process for photons in a medium with nonlinear susceptibility. The process is very general since it involves an arbitrary number of quanta in an arbitrary number of modes. It is special because of the restriction to detailed balance, which amounts here to the assumption of equal loss rates for all modes. In Section 8.2 the Fokker-Planck equation for the process is set up along the lines given in Section 5.2. This equation is solved in 8.3 by using the methods of Section 4.3. In 8.4 we discuss some special examples contained in the general solution. Some of these cases were already considered in Section 7 from a purely phenomenological viewpoint. As a new result we obtain the stationary distributions of higher order parametric processes. Furthermore, multi-mode effects, both in the pump and in the stimulated processes are taken into account.

8.1. A Class of Scattering Processes in Nonlinear Optics and Detailed Balance

We consider optical scattering processes of the following kind. Let some medium with a field dependent optical susceptibility be given, in which certain optical modes can propagate. One mode is supposed to be directly excited by an external laser field. The quanta which are present in this directly excited mode may be scattered into other modes. The interaction which causes this scattering is mediated by the electrons of the medium, i.e., by the field dependent part of its susceptibility. We represent this process graphically in Fig. 19. A single photon of a field mode (with frequency ω_p and wavenumber k_p) which is directly coupled to the pump light, decays into n photons with various frequencies and wavenumbers. The conservation of energy and momentum implies the matching conditions

$$\sum_{\nu} n_\nu \omega_\nu = \omega_p , \tag{8.1}$$

$$\sum_{\nu} n_\nu k_\nu = k_p , \tag{8.2}$$

where we have allowed for the creation of n_ν quanta in mode ν. If we assume that there is no resonance between electronic transitions in the medium and the frequencies of the modes, the interaction between the modes may be described by an effective Hamiltonian (cf. e.g. [9, 44, 77, 79]). Attaching a boson annihilation operator b_ν to each incoming line in Fig. 19 and a boson creating operator b_ν^+ to each outgoing line, we get the following effective interaction Hamiltonian

$$H_{\text{int}} = (ih\, F_p b_p^+ + ih\gamma\, b_p \Omega^+) + (\text{h.c.}) \tag{8.3}$$

with

$$\Omega^+ = \prod_{\nu=1}^{n} (b_\nu^+)^{n_\nu}. \tag{8.4}$$

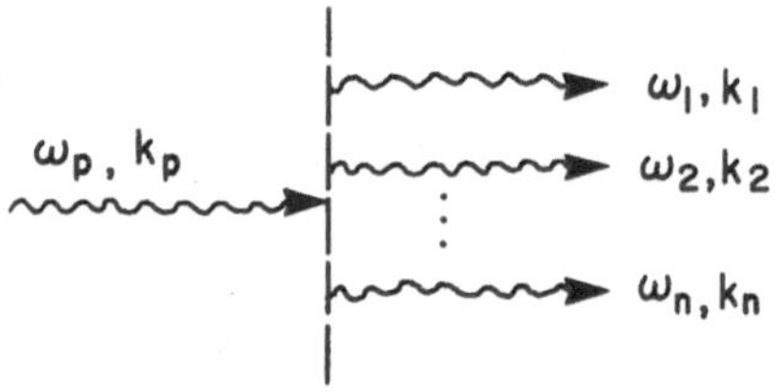

Fig. 19. General scattering process in nonlinear optics

The first term in Eq. (8.3) describes the direct excitation of the mode with frequency ω_p by the external force F_p, which is proportional to the amplitude of the pumping laser. The second term describes the scattering process shown in Fig. 19, where we have again allowed for the simultaneous creation of several (n_ν) quanta in mode ν. The hermitian conjugate describes the time reversed processes. We may generalize the Hamiltonian (8.3) still further by allowing for the presence of several different competing scattering mechanisms of the type shown in Fig. 19. We simply have to put, instead of Eq. (8.4),

$$\Omega^+ = \Omega^*(b_1^+, b_2^+, \ldots, b_n^+) \tag{8.5}$$

where Ω^* is some arbitrary (analytic) function, defined for c-numbers. The resonance condition (8.1) has to be generalized to the condition

$$\begin{aligned} &\Omega^*(b_1^+ \exp i\omega_1 t, b_2^+ \exp i\omega_2 t, \ldots, b_n^+ \exp i\omega_n t) \\ &= \exp i\omega_p t \cdot \Omega^*(b_1^+, b_2^+, \ldots, b_n^+). \end{aligned} \tag{8.6}$$

We assume that this scattering process occurs between the two mirrors of some optical cavity. The mirror losses of all modes have to be taken into account by additional terms in the Hamiltonian, which describe the coupling of the mode amplitudes to some heat baths. Eliminating the heat bath variables by well known procedures, we derive an equation of motion for the density operator $\varrho(b_1, b_1^+, b_2, b_2^+, \ldots b_n, b_n^+, b_p, b_p^+)$ (see, e.g., [82, 77]). ϱ may be considered to depend on the basic boson operators describing the field modes.

$$\begin{aligned}\dot{\varrho} = -i\hbar^{-1}[H_{\text{int}}, \varrho] + \sum_{\nu} \{ &\kappa_\nu(\bar{n}_\nu + 1)([b_\nu \varrho, b_\nu^+] + [b_\nu, \varrho b_\nu^+]) \\ &+ \kappa_\nu \bar{n}_\nu([b_\nu^+ \varrho, b_\nu] + [b_\nu^+, \varrho b_\nu]) \\ &+ \kappa_p(\bar{n}_p + 1)([b_p \varrho, b_p^+] + [b_p, \varrho b_p^+]) \\ &+ \kappa_p(\bar{n}_p + 1)([b_p^+ \varrho, b_p] + [b_p^+, \varrho b_p])\} .\end{aligned} \tag{8.7}$$

Here the constant κ_ν is the decay rate of the amplitude of mode ν due to the escape of photons through the mirrors of the cavity. $\bar{n}_\nu$ is the mean quantum number in mode ν due to the presence of the heat baths alone. For thermal heat baths and optical frequencies this number is much smaller than 1 and therefore neglected. The notation κ_p and $\bar{n}_p$ is evident; $\bar{n}_p$ is also negligible. Eq. (8.7) gives the complete microscopic formulation of our problem. Due to the presence of the driving force F_p and the loss rates κ_ν, κ_p in (8.7), we have a steady energy flow from the pumping laser through the directly excited mode into the other modes. The distribution of the energy over the different modes and their degree of excitation is determined by their loss rates and their participation in the scattering process. Due to the presence of mirrors and feedback, multiple scattering processes and depletion of the initially excited mode are important. At first sight, this problem seems to be very complicated. On the other hand, it is clear from the discussion in section 4 that it is possible to find the stationary distribution for systems in detailed balance, even for very complicated cases. Hence, we reduce our general problem to a special one, in which we may expect the presence of detailed balance. This could be done by deriving a Fokker-Planck equation from Eq. (8.7) and looking for cases in which the potential conditions of 4.2 are fulfilled.

Physically more instructive, although mathematically less rigorous, is the method of directly analyzing the possibility of irreversible circular probability currents in the system. The parameters of the system can then be chosen in such a way as to make circular probability currents impossible. Here we apply the latter procedure. In the present case, circular probability currents may occur in two different ways: first, by a separate circular diffusion of the amplitude of each mode; second, by a coupled circular motion of several mode amplitudes. It can be shown by the arguments of Section 6.1 that the probability currents due to uncoupled amplitude motions always have to be reversible and cannot destroy detailed balance. Thus, the only possibility for circular probability currents is the occurrence of circular motions involving several amplitudes. Fig. 20 shows the way in which the variables of the total system (including the pumping laser and the heat baths) are coupled by the Hamiltonian. This scheme indicates which variables can participate in a coupled cyclic diffusion. Since the total system is not in thermal equilibrium, some cyclic probability currents have to occur.

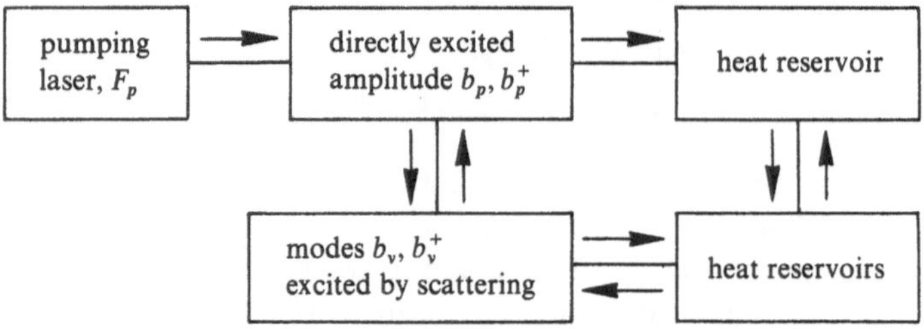

Fig. 20. The coupling of the various degrees of freedom in the total system

We consider first the directly excited mode, described by b_p. This degree of freedom participates in a cyclic current, because it is excited by the laser field without having the possibility of acting back on the laser. However, if we make the assumption, that this degree of freedom relaxes rapidly to a conditional equilibrium with respect to the amplitudes of the other modes, i.e.,

$$\kappa_p \gg \kappa_\nu \tag{8.8}$$

we can eliminate this variable without destroying the Markoff property of the remaining mode amplitudes. Then it is sufficient that the irreversible cyclic probability currents among the remaining degrees of freedom vanish, in order to have detailed balance. Fig. 21 incidates how the parametrically excited modes are coupled. They are indirectly coupled by the directly excited mode and they are also coupled by the heat baths.

In principle, we may have a circular probability current in which two or more modes participate in the following way: The amplitude of mode μ is changed by a fluctuation absorbed by its heat bath, which is at the same temperature as the heat bath of mode ν. The heat bath of mode ν transfers the fluctuation into mode ν, which reacts back on the amplitude b_p. From there, the fluctuation is given to mode μ again, thereby closing the cycle.

From Fig. 21 it becomes clear, that circular probability currents of this type are zero, if the net rate with which the quanta of different modes are absorbed by the heat baths are equal for all modes. Therefore, detailed balance holds if

$$\kappa_\nu = \kappa\,. \tag{8.9}$$

If we assume Eq. (8.9) we can find the stationary solution by the methods of 4.3. But first, we have to derive a Fokker-Planck equation from Eq. (8.7).

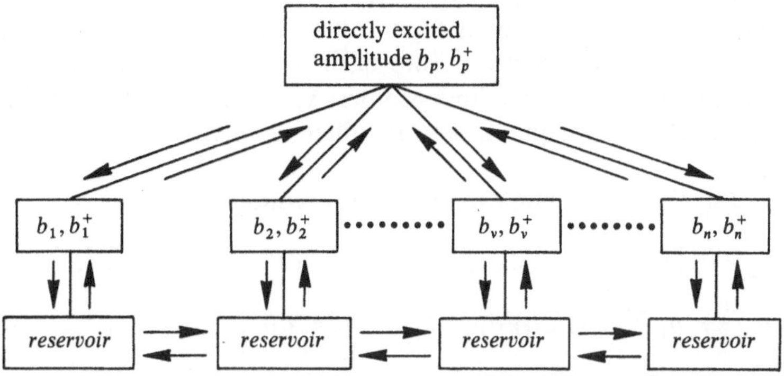

Fig. 21. The coupling of the parametrically excited modes by the heat reservoirs and the directly excited mode

8.2. Fokker-Planck Equations for the *P*-Representation and the Wigner Distribution

We transform the operator equation (8.7) into an equivalent *c*-number equation by using a quasi-probability representation (cf. [8, 84]) of the statistical operator ϱ. We obtain the *P*-representation [85] of ϱ,

$$\varrho(\{b, b^+\}) = (\pi)^{-n} \prod_{(\nu)} \int d^2\beta_\nu |\{\beta\}\rangle\, P(\{\beta, \beta^*\})\, \langle\{\beta\}|\,, \tag{8.10}$$

by putting ϱ into anti-normal order with respect to the boson operators of the modes and considering it as a function of $\{\beta, \beta^*\}$ instead of $\{b, b^+\}$ [86]. In Eq. (8.10), $|\{\beta\}\rangle$ is a coherent state [84], defined by

$$b_\nu |\{\beta\}\rangle = \beta_\nu |\{\beta\}\rangle\,. \tag{8.11}$$

The equation of motion for *P* is obtained [86] by putting the operator equation of motion for ϱ, (8.7), into anti-normal order and substituting

$$b_\nu \to \beta_\nu\,, \quad b_\nu^+ \to \beta_\nu^+\,. \tag{8.12}$$

The ordering can be carried out with the help of the relations

$$\varrho b_\nu = (b_\nu - (\partial/\partial b_\nu^+))\,\varrho\,, \tag{8.13}$$

$$\varrho b_\nu^+ = (b_\nu^+ + (\partial/\partial b_\nu))\,\varrho\,. \tag{8.14}$$

We obtain

$$\dot{P} = \Big[\kappa \sum_\nu \partial/\partial\beta_\nu\, \beta_\nu P + \gamma\,(\beta_p \Omega^*(\{\beta^* - \partial/\partial\beta\}) - (\beta_p - \partial/\partial\beta_p^*)\, \Omega^*(\{\beta^*\}))\, P$$
$$- F_p^*\, \partial/\partial\beta_p^*\, P + \kappa_p\, \partial/\partial\beta_p^*\, \beta_p^*\, P\Big] + [\text{c.c.}]\,. \qquad (8.15)$$

Eq. (8.15) can be approximated by a Fokker-Planck equation by expanding the differential operator $\Omega^*(\{\beta^* - \partial/\partial\beta\})$ in powers of $\partial/\partial\beta$ and retaining only terms up to second order. One obtains essentially an expansion around the classical limit of the considered quantum process (cf. 5.2). By introducing the dimensionless variables $\hat{\beta}_\nu = \beta_\nu/\sqrt{\langle|\beta_\nu|^2\rangle}$ one can show, that subsequent powers differ in order of magnitude by $1/\langle|\beta_\nu|^2\rangle$.

In this diffusion approximation Eq. (8.15) reduces to

$$\dot{P} = \Big[\sum_\nu \partial/\partial\beta_\nu^*(\kappa\beta_\nu^* - \gamma\beta_p^*\, \partial\Omega(\{\beta\})/\partial\beta_\nu) P + \partial/\partial\beta_p^*(\kappa_p\beta_p^* - F_p^* + \gamma\Omega^*(\{\beta^*\}))P$$
$$+ \tfrac{1}{2}\gamma \sum_{\nu\mu} \beta_p\, \partial^2\Omega^*(\{\beta^*\})/\partial\beta_\nu^*\, \partial\beta_\mu^*\, \partial^2 P/\partial\beta_\nu\, \partial\beta_\mu\Big] + [\text{c.c.}]\,. \qquad (8.16)$$

Once the solution of Eq. (8.16) is known, all normally ordered expectation values of the boson operators may be calculated by substituting according to Eq. (8.12) and using the distribution like a classical probability density [85]. A disadvantage of the use of the P-representation is that P need not always exist. In particular it is known that P does not always exist for parametric processes involving several modes [75, 77]. For this reason we prefer to transform Eq. (8.15) into an equation for the Wigner distribution, which is known always to exist [87, 84]. The Wigner distribution is given in terms of P by

$$W(\{\beta, \beta^*\}) = (2/\pi)^n \prod_{(\nu)} \int d^2\alpha_\nu (\exp - 2|\alpha_\nu - \beta_\nu|^2)\, P(\{\alpha, \alpha^*\})\,. \qquad (8.17)$$

From Eq. (8.15) we obtain

$$\dot{W} = \Big[\kappa \sum_\nu \partial/\partial\beta_\nu^*(\beta_\nu^* + \tfrac{1}{2}\partial/\partial\beta_\nu)\, W + \gamma(\beta_p + \tfrac{1}{2}\partial/\partial\beta_p^*)\, \Omega^*(\{\beta^* - \tfrac{1}{2}\partial/\partial\beta\})\, W$$
$$- \gamma(\beta_p - \tfrac{1}{2}\partial/\partial\beta_p^*)\, \Omega^*(\{\beta^* + \tfrac{1}{2}\partial/\partial\beta\})\, W - F_p^*\, \partial W/\partial\beta_p^* \qquad (8.18)$$
$$+ \kappa_p\, \partial/\partial\beta_p^*(\beta_p^* + \tfrac{1}{2}\partial/\partial\beta_p)\, W\Big] + [\text{c.c.}]\,.$$

As before, we introduce the diffusion approximation and obtain

$$\dot{W} = \Big[\sum_\nu \partial/\partial\beta_\nu^*(\kappa\beta_\nu^* - \gamma\beta_p^*\, \partial\Omega(\{\beta\})/\partial\beta_\nu)\, W + \sum_\nu \tfrac{1}{2}\kappa\, \partial^2 W/\partial\beta_\nu\, \partial\beta_\nu^*$$
$$+ \partial/\partial\beta_p^*(\kappa_p\beta_p^* - F_p^* + \gamma\Omega^*(\{\beta^*\}))\, W + \tfrac{1}{2}\kappa_p\, \partial^2 W/\partial\beta_p\, \partial\beta_p^*\Big] \qquad (8.19)$$
$$+ [\text{c.c.}]\,.$$

It is interesting to compare Eqs. (8.19) and (8.16). They differ in the second order derivative terms which describe the fluctuations, i.e. the spontaneous emission processes. The two different descriptions correspond to two different, but equivalent, interpretations of spontaneous emission in nonlinear optics. Spontaneous emission can be considered as being induced by the vacuum fluctuations of the modes, which are in turn driven by the vacuum fluctuations of the reservoirs ($\sim\kappa$). Spontaneous emission may also be viewed as arising from the interaction ($\sim\gamma$).

If W has been determined from Eq. (8.19), one may calculate all normally ordered expectation values by substituting

$$b_\nu^+ \to \beta_\nu^* + \tfrac{1}{2}\,\partial/\partial\beta_\nu\,; \qquad b_\nu \to \beta_\nu + \tfrac{1}{2}/\partial\beta_\nu^* \tag{8.20}$$

and using W as if it were a classical probability density. Antinormally ordered expectation values can be evaluated by substituting

$$b_\nu^+ \to \beta_\nu^* - \tfrac{1}{2}\,\partial/\partial\beta_\nu\,; \qquad b_\nu \to \beta_\nu - \tfrac{1}{2}\,\partial/\partial\beta_\nu^* \tag{8.21}$$

and proceeding as before (cf. [88]).

8.3. Stationary Distribution for the General Process

Neither Eq. (8.16) nor Eq. (8.19) fulfill the potential conditions of Section 4.2. The reason for this was discussed in 8.1 and was found to be given by certain cyclic probability currents occurring in the stationary state. Adopting now the two conditions (8.8) and (8.9), we can suppress these currents, and hence establish detailed balance. Condition (8.8) is used to determine the equilibrium value of the amplitude β_p of the directly excited mode, for given values of the pump F_p and the instantaneous amplitudes of the remaining modes. Putting the drift term of β_p in Eq. (8.19) equal to zero we obtain

$$\beta_p = F_p/\kappa_p - \gamma\Omega/\kappa_p\,. \tag{8.22}$$

We eliminate β_p from Eq. (8.19) by inserting Eq. (8.22) and integrating over β_p, β_p^*. In addition, we use the condition of equal damping, Eq. (8.9). This gives us the reduced equation

$$\begin{aligned}\dot W = \sum_\nu \{ &\partial/\partial\beta_\nu^*[\kappa\beta_\nu^* - (\gamma F_p^*/\kappa_p)\,\partial\Omega/\partial\beta_\nu + (\gamma^2/\kappa_p)\,\partial|\Omega|^2/\partial\beta_\nu]\,W\\ &+ \partial/\partial\beta_\nu[\kappa\beta_\nu - (\gamma F_p/\kappa_p)\,\partial\Omega/\partial\beta_\nu^* + (\gamma^2/\kappa_p)\,\partial|\Omega|^2/\partial\beta_\nu^*]\,W\\ &+ \kappa\,\partial^2 W/\partial\beta_\nu\partial\beta_\nu^*\}\,.\end{aligned} \tag{8.23}$$

Eq. (8.23) fulfills the potential conditions as it should according to the analysis of Section 8.1. It is now easy to give the stationary solution of Eq. (8.23) by making use of the connection between the drift term and the stationary distribution, which is provided by the potential condition (4.13). We obtain

$$W_1^s \sim \exp(-\phi^s) \tag{8.24}$$

with

$$\phi^s = 2\sum_{\nu} |\beta_\nu|^2 - (2\gamma/\kappa\kappa_p)(F_p^*\Omega + F_p\Omega^*) + (2\gamma^2/\kappa\kappa_p)|\Omega|^2 . \tag{8.25}$$

Eq. (8.25) gives the stationary distribution directly in terms of the arbitrary function Ω, which defines the interaction Hamiltonian (8.3). Therefore, this solution is very general and comprises many different special cases (cf. [80, 89]). For $\gamma = 0$, the result (8.24), (8.25) reduces to the Wigner distribution of independent modes in their vacuum state. For $F_p = 0$, Eqs. (8.24), (8.25) describe modes, which are passively coupled by the nonlinear properties of the medium. The same result would be obtained, if F_p is different from zero but fluctuates on a very short time scale. This result explains the need of coherent laser sources for pumping oscillators in nonlinear optics. In the following section we consider a number of special cases contained in the solution (8.15).

8.4. Examples

The most important scattering processes in nonlinear optics are those in which quanta in two modes are created. These second order effects usually have the largest cross sections among the nonlinear processes. These processes were already considered in Section 7 from the viewpoint of the phenomenological theory. We can now compare these results with the results of the present microscopic theory. Moreover, we discuss the influence of various kinds of multimode effects on the photon statistics. These effects may arise from the multimode structure of the pump. They can also come from the multimode structure of the output, which may contain several signal-idler clusters. Higher order scattering processes, which include three or more scattered quanta created by the destruction of one initial quantum, have a considerably smaller cross section and are therefore more difficult to observe in practice. Some of their properties, discussed in section *b*, differ quite markedly from the properties predicted for second order effects. In particular it is shown, that the instability leading to oscillation in such higher order modes is not a continuous instability, like that of the second order effects, which resembles continuous symmetry breaking second order phase transition (cf. 5.4). Instead, we find that the oscillation threshold is marked by a discontinuous jump of the mode amplitudes from zero to some finite value. The instability causes a symmetry change as in the earlier examples.

Thus, the oscillation threshold of these higher order processes resembles a first order phase transition. This difference in the threshold behaviour should be observable experimentally and would make the experimental investigation of these processes worthwhile.

a) Parametric and Subharmonic Oscillation

i) Parametric Oscillation

For the case of parametric oscillation the basic scattering process is represented by Fig. 22. The number of modes is $n=2$. Ω takes the form

$$\Omega^{+}=\Omega^{*}(b_1^{+},b_2^{+})=b_1^{+}\,b_2^{+}\,. \tag{8.26}$$

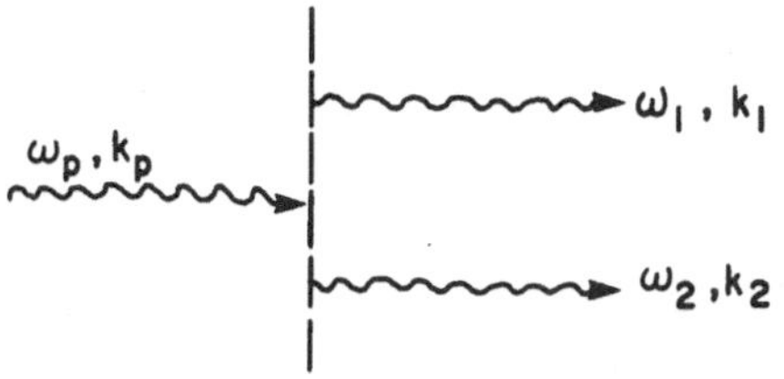

Fig. 22. Parametric scattering

Therefore, the potential ϕ^s is given, according to Eq. (8.25), by

$$\phi^s=2|\beta_1|^2+2|\beta_2|^2-(2\gamma/\kappa\kappa_p)(F_p^*\beta_1\beta_2+F_p\beta_1^*\beta_2^*)+(2\gamma^2/\kappa\kappa_p)|\beta_1\beta_2|^2\,. \tag{8.27}$$

This result has already been obtained in [80]. It was compared with the result of the phenomenological theory in Section 7.1. The potential (8.27) has the continuous symmetry

$$\Delta\varphi=\varphi_1-\varphi_2\rightarrow\Delta\varphi'=\varphi_1'-\varphi_2' \tag{8.28}$$

with

$$\varphi_1+\varphi_2=\varphi_1'+\varphi_2'\,. \tag{8.29}$$

Here, φ_1 and φ_2 are the phases of the complex amplitudes β_1, β_2. For

$$|F_p|\leqq\kappa\kappa_p/\gamma\,, \tag{8.30}$$

the potential ϕ^s has a minimum for $\beta_1=\beta_2=0$. This minimum has the full symmetry (8.28) and is, therefore, not degenerate. For

$$|F_p|>\kappa\kappa_p/\gamma \tag{8.31}$$

ϕ^s has a continuum of degenerate minima, given by

$$|\beta_1| = |\beta_2| = \sqrt{|F_p|/\gamma - \kappa\kappa_p/\gamma^2}\,. \tag{8.32}$$

Each single minimum (8.32) breaks the symmetry (8.28) which is the reason for the degeneracy. Assuming sharp values for the amplitudes $|\beta_1|$ and $|\beta_2|$ well above threshold, we obtain from Eq. (8.27)

$$W_1^s \sim \exp[(4\gamma|F_p^* \beta_1 \beta_2|/\kappa\kappa_p) \cdot \cos(\varphi_1 + \varphi_2 - \psi)] \tag{8.33}$$

where ψ is the phase of F_p. Eq. (8.33) is the same expression as obtained in the case of mode locking in Section 6.3. It demonstrates that the sum $\varphi_1 + \varphi_2$ is, indeed, locked to the phase of the pumping laser. Integrating Eq. (8.27) over φ_1 and φ_2 we obtain the distribution for the amplitudes $|\beta_1|$ and $|\beta_2|$[15].

$$W_1^s \sim |\beta_1 \beta_2|\, I_0(4\gamma|F_p \beta_1 \beta_2|/\kappa\kappa_p) \exp[-2(|\beta_1|^2 + |\beta_2|^2 + \gamma^2 |\beta_1 \beta_2|^2/\kappa\kappa_p)]\,. \tag{8.34}$$

I_0 is the Bessel function with imaginary argument and index 0. For typical oscillators the quantity γ is much smaller than κ. Hence,

$$\kappa\kappa_p/\gamma^2 \gg 1\,. \tag{8.35}$$

With (8.35) and the notation

$$I_{1,2} = 2|\beta_{1,2}|^2/\alpha_0\,, \tag{8.36}$$

$$a = 4(|F_p|/\sqrt{2\kappa\kappa_p} - \alpha_0)\,, \tag{8.37}$$

$$\alpha_0 = \sqrt{\kappa\kappa_p/2\gamma^2}\,, \tag{8.38}$$

Eq. (8.34) takes on the more transparent form

$$W_1^s \sim (I_1 I_2)^{-1/4} \exp[(\tfrac{1}{2} a \sqrt{I_1 I_2} - \tfrac{1}{4} I_1 I_2) - \alpha_0(\sqrt{I_1} - \sqrt{I_2})^2]\,. \tag{8.39}$$

This distribution is shown in Fig. 23. Because of (8.35), we have $\alpha_0 \gg 1$, and the distribution is sharply centered around $\sqrt{I_1} = \sqrt{I_2}$. Therefore, the assumption of the phenomenological theory, that only one degree of freedom becomes unstable at threshold (cf. Section 3.2 for the general case and Section 7.1 for the special example), is clearly born out by the microscopic result. Integrating Eq. (8.34) over the idler amplitude $|\beta_2^2|$ we obtain

$$W_1^s \sim \frac{|\beta_1|}{1 + |\beta_1|^2/2\alpha_0^2} \exp\left\{-2|\beta_1|^2 + \frac{(a + 2\alpha_0)|\beta_1|^2}{\alpha_0(1 + |\beta_1|^2/2\alpha_0^2)}\right\} \tag{8.40}$$

which reduces to the result (7.19) for $\alpha_0 \gg 1$.

[15] The normalization has to be carried out by integrating over $|\beta_{1,2}|$ from 0 to ∞.

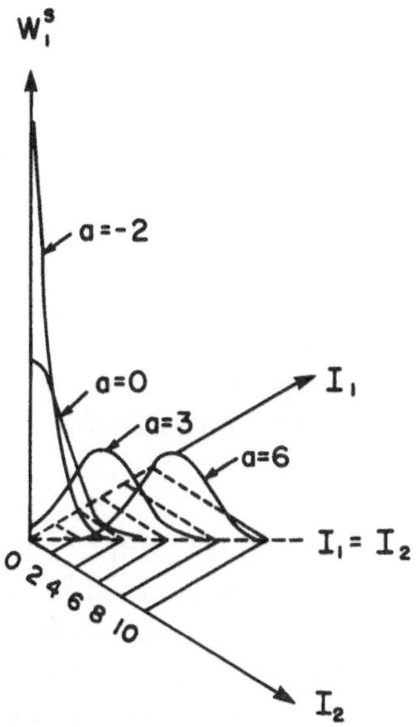

Fig. 23. The stationary distribution of the intensities I_1, I_2 of signal and idler, Eq. (8.39) for various pump intensities a, Eq. (8.37). The sharp concentration of the distribution around $I_1 = I_2$ is not resolved in the diagram because of Eq. (8.35)

ii) Subharmonic Oscillation

Experimentally, subharmonic oscillation is realized, if the two modes cannot be distinguished from each other by any method. For this case the basic scattering process is shown in Fig. 24. The number of modes is $n=1$. Ω takes the form

$$\Omega^+ = \Omega^*(b_1^+) = b_1^{+\,2}. \tag{8.41}$$

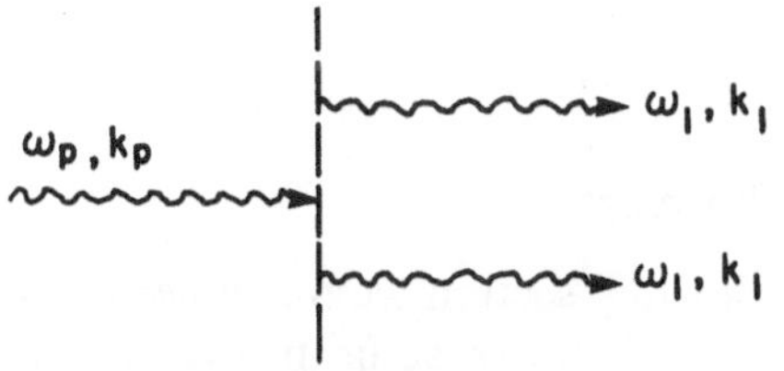

Fig. 24. Subharmonic scattering

It should be noted, that the distribution for the subharmonic amplitude is not obtained as the limiting case of the signal distribution (8.40), but rather as the limiting case of the joint distribution of signal and idler Eq. (8.27). The potential ϕ^s is given, according to Eq. (8.25), by

$$\phi^s = 2|\beta_1|^2 - (2\gamma/\kappa\kappa_p)(F_p^*\beta_1^{*2}) + (2\gamma^2/\kappa\kappa_p)|\beta_1|^4. \tag{8.42}$$

This result coincides in form with the result of the phenomenological theory, Eq. (7.27), and a result obtained in [82]. The theory given in [82] was purely classical and dealt with thermal fluctuations in subharmonic generation. Contrary to that treatment, we neglected in 8.1 all thermal fluctuations as being unimportant at optical frequencies, and considered, instead, the quantum fluctuations as constituting the essential noise source. The fact, that both theories give the same overall result, apart from different expressions for the parameters of the distribution, is immediately understood in view of the purely phenomenological arguments given in 7.2. The potential (8.42) is shown in Fig. 18. The minimum at $\beta_1 = \beta_2 = 0$ becomes unstable for

$$|F_p| > \kappa\kappa_p/2\gamma . \tag{8.43}$$

At first sight it seems that the threshold for subharmonic generation is half the threshold for nondegenerate parametric oscillation (cf. Eq. (8.31)). However, it should be noted that κ in Eq. (8.43) is the limit of the sum of the loss rates of the nondegenerate modes, i.e., $\kappa_1 + \kappa_2 \to \kappa$. Therefore, the 2 in (8.43) cancels. The new minima of ϕ^s are given by

$$\beta_1 = \beta_1^* = \pm \sqrt{|F_p|/\gamma - \kappa\kappa_p/2\gamma^2} . \tag{8.44}$$

Since they break a discontinuous symmetry, there exists no continuous motion, which could restore the symmetry. Instead, the symmetry is restored by discontinuous jumps between the two minima. The mean time τ between two such jumps has been calculated in [82] for thermal fluctuations. For quantum fluctuations this result takes the form

$$\tau = \frac{\kappa_p \sqrt{1 + \kappa\kappa_p/2\gamma F_p}}{4\pi(2\gamma F_p - \kappa\kappa_p)} \exp\left(\frac{(2\gamma F_p - \kappa\kappa_p)^2}{\gamma^2 \kappa\kappa_p}\right) \tag{8.45}$$

Eq. (8.45) is valid for

$$1 > (2\gamma F_p - \kappa\kappa_p)/\kappa\kappa_p > \sqrt{2\kappa/\kappa_p} . \tag{8.46}$$

b) *Higher Order Processes*

In this, and the following section we make use of the solution (8.25) to discuss the photon statistics of some processes, which have not been discussed earlier in the literature.

We consider first the n-photon process, which is represented by Fig. 19 and choose Ω^+ according to Eq. (8.4). The general solution (8.25) then takes the form

$$\begin{aligned} \phi^s = 2\sum_\nu |\beta_\nu|^2 - (2\gamma/\kappa\kappa_p)\left(F_p^* \prod_{(\nu)} (\beta_\nu)^{n_\nu} + F_p \prod_{(\nu)} (\beta_\nu^*)^{n_\nu}\right) \\ + (2\gamma^2/\kappa\kappa_p) \prod_{(\nu)} |\beta|^{2n_\nu} . \end{aligned} \tag{8.47}$$

We split the complex amplitudes into absolute values and phases $\beta_\nu = r_\nu \exp - i\varphi_\nu$.
The potential ϕ^s has, in general, a large number of continuous symmetries, since all phases φ_ν may be changed continuously, subject to the constraint

$$\sum_\nu n_\nu \varphi_\nu = \text{const}. \tag{8.48}$$

The extrema of (8.47) are determined by the equations

$$0 = r_\nu - (n_\nu \gamma/\kappa\kappa_p r_\nu)\, F_p \prod_{(\mu)} (r_\mu)^{n_\mu} + (n_\nu \gamma^2/\kappa\kappa_p r_\nu) \prod_{(\mu)} (r_\mu)^{2n_\mu} \tag{8.49}$$

for $\nu = 1 \dots n$. Without restriction of generality we have chosen the phase of F_p to be zero. By the transformation

$$r_\nu = \sqrt{n_\nu}\, (\kappa\kappa_p/\gamma^2 \alpha)^{1/(2\sum_\mu n_\mu - 2)} \cdot \hat{r}_\nu \tag{8.50}$$

with

$$\alpha \equiv \prod_{(\nu)} n_\nu^{n_\nu}, \tag{8.51}$$

we obtain the equations

$$0 = \hat{r}_\lambda^2 - F_p(\kappa\kappa_p)^{-1/2} (\gamma^2 \alpha/\kappa\kappa_p)^{1/(2\sum_\mu n_\mu - 2)} \prod_{(\nu)} (\hat{r}_\nu)^{n_\nu} + \prod_{(\nu)} (\hat{r}_\nu)^{2n_\nu} \tag{8.52}$$

for $\lambda = 1 \dots n$, whose coefficients are completely independent of the index λ. Therefore, the system of Eq. (8.52) is solved by putting

$$\hat{r}_\nu = r \tag{8.53}$$

independent of ν.
Eq. (8.50) gives directly the relative magnitude of the absolute values of the various mode amplitudes. For r we obtain the trivial solution

$$r = 0 \tag{8.54}$$

and, in addition

$$\begin{aligned} &-1 + F_p(\kappa\kappa_p)^{-1/2} (\gamma^2 \alpha/\kappa\kappa_p)^{1/(2\sum_\mu n_\mu - 2)} (r)^{-2 + \sum_\mu n_\mu} \\ &= (r)^{-2 + 2\sum_\mu n_\mu} \end{aligned} \tag{8.55}$$

We observe, that the right hand side of Eq. (8.55) is formed by a power of r of the order $2 \sum_\nu n_\nu - 2$, which passes through 0 for $r = 0$ (cf. Fig. 26). The left hand side is formed by a polynomial of the lower order $\sum_\nu n_\nu - 2$, whose coefficient is proportional to F_p. It goes through -1 for $r = 0$. For $F_p = 0$ and for sufficiently small F_p, the left hand side is always smaller than the right hand side and no real solution of Eq. (8.55) exists.

In this case $r=0$ is the only solution and we have a stable behaviour with small amplitudes. If F_p becomes sufficiently large, $F_p=F_{p_c}$, the polynomial on the left hand side will finally, at some point, be tangent to the power of the right hand. For still larger values of F_p, the two curves will intersect in two points. Exemptions to this general behaviour are all processes with $\sum_\nu n_\nu \leqq 2$, which are the second order processes

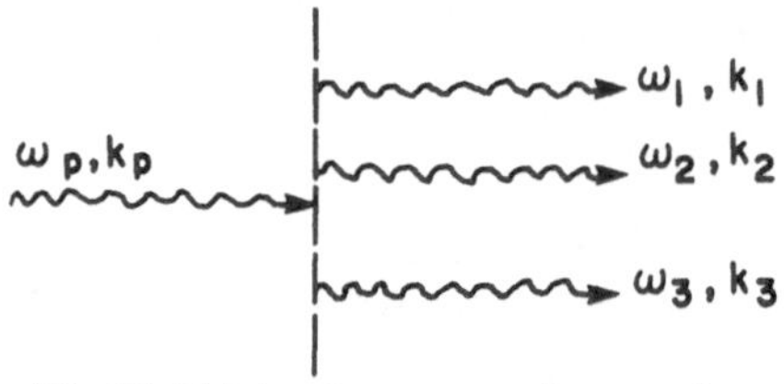

Fig. 25. Third order parametric scattering

discussed in *a*). The first intersection for smaller r defines a maximum of ϕ^s which corresponds to an instable state. The second intersection for larger r gives a new set of degenerate minima of ϕ^s, which correspond to new stable stationary states, each of which breaks the phase symmetries. Since, in general, several (continuous) phase symmetries are present, the new minima are degenerate with respect to several (continuous) parameters. Therefore, several different diffusion modes exist which carry the system through the degenerate new minima. With increasing F_p the maximum of ϕ^s is shifting to smaller values of r whereas the new minimum shifts to larger r values. We find, therefore, a multistable behaviour with the two stable states at $r=0$ and at $r>0$. We consider several special cases of Eq. (8.55).

i) For the scattering process Fig. 25, we obtain

$$-1+|F_p|\sqrt{\gamma}(\kappa\kappa_p)^{-3/4}\,r=r^4\,. \tag{8.56}$$

The three cases $F_p \lesseqgtr F_{p_c}$ are shown in Fig. 26.

ii) We consider the case in which the three modes of Fig. 25 degenerate to a single mode, the third order subharmonic with $\omega=\frac{1}{3}\omega_p$. In this case Eq. (8.47) gives a potential with the discontinuous threefold symmetry

$$\varphi\to\varphi+\tfrac{2}{3}\pi\to\varphi+\tfrac{4}{3}\pi\,. \tag{8.57}$$

This potential is shown in Fig. 27. The minimum at $r=0$ has the full symmetry (8.57). For F_p larger than some critical value F_{p_c}, the equation

$$-1+27^{1/4}|F_p|\gamma^{1/2}(\kappa\kappa_p)^{-3/4}\,r=r^4 \tag{8.58}$$

which follows from Eq. (8.55), gives a new minimum of ϕ^s for some root $r>0$ (cf. Fig. 26). This new minimum is degenerate with two further minima, according to the symmetry (8.57).

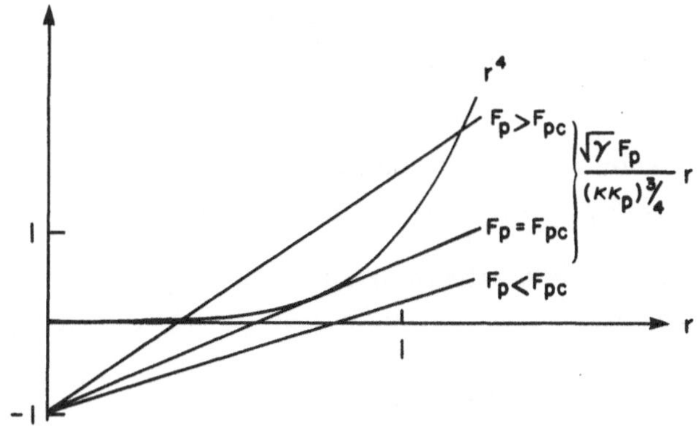

Fig. 26. Threshold condition and extrema of the potential ϕ^s for third subharmonic generation

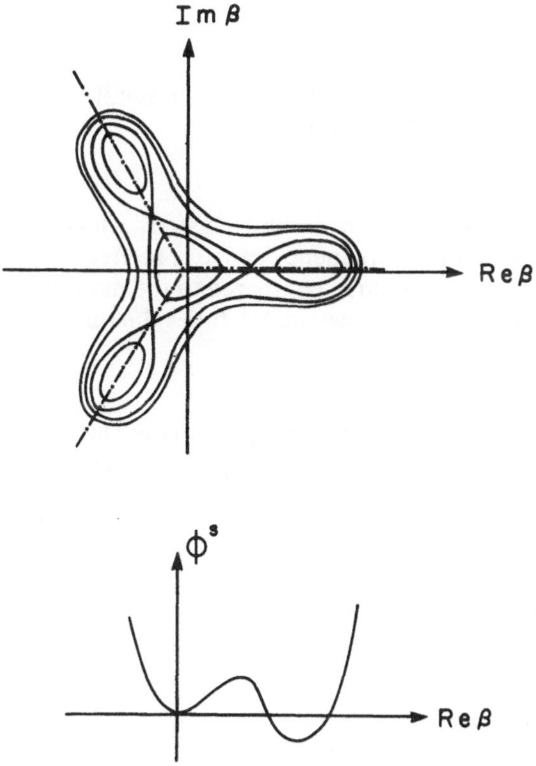

Fig. 27. The potential (8.47) for third subharmonic generation above threshold

iii) For the general case of the m-th subharmonic, Eq. (8.47) has an m-fold discontinuous symmetry

$$\varphi \to \varphi + 2\pi/m \to \cdots \to \varphi + 2\pi k/m \to \cdots \to \varphi + 2\pi . \tag{8.59}$$

The threshold condition is determined by

$$-1 + F_p(\kappa\kappa_p)^{-1/2} (\gamma^2 m^m/\kappa\kappa_p)^{1/(2m-2)} r^{m-2} = r^{2m-2} . \tag{8.60}$$

All that has been said for the third subharmonic can be carried over to this more general case.

c) Mode Clusters in Parametric Oscillators

We consider now an example in which Ω, in the Hamiltonian (8.3), describes several competing scattering processes – the simultaneous oscillation of several signal-idler clusters in parametric oscillators. A cluster is defined as a pair consisting of one signal and one idler. Several clusters may oscillate simultaneously for the following well known reason [81]: The matching condition

$$k_1 + k_2 = k_p \tag{8.61}$$

has, in oscillators of finite length, only to be fulfilled up to multiples of $2\pi/L$. This weaker condition, together with

$$\omega_1 + \omega_2 = \omega_p \tag{8.62}$$

is satisfied simultaneously by several clusters, which are equidistantly distributed over a broad frequency range. The frequency difference between adjacent clusters is large compared to the natural mode spacing in the resonator cavity. For l clusters, Ω^+ takes the form

$$\Omega^+ = \sum_\nu g_\nu b_{1\nu}^+ b_{2\nu}^+ \tag{8.63}$$

where $b_{1\nu}^+$ and $b_{2\nu}^+$ are the creation operators of the signal and the idler in the ν-th cluster. The general potential (8.25) then takes the form

$$\begin{aligned} \phi^s = 2\sum_\nu (|\beta_{1\nu}|^2 + |\beta_{2\nu}|^2) - 2\gamma(\kappa\kappa_p)^{-1} F_p \sum_\nu (g_\nu \beta_{1\nu}^* \beta_{2\nu}^* + \text{c.c.}) \\ + 2\gamma^2(\kappa\kappa_p)^{-1} \sum_{\nu\nu'} g_\nu g_{\nu'}^* \beta_{1\nu}^* \beta_{1\nu'} \beta_{2\nu}^* \beta_{2\nu'} \end{aligned} \tag{8.64}$$

where the phase of F_p has been put equal to zero. For

$$F_p \leqq \kappa\kappa_p/\gamma|g_\nu| \tag{8.65}$$

ϕ^s has a minimum at $\beta_\nu = 0$. For

$$F_p > \kappa\kappa_p/\gamma|g_\nu| \tag{8.66}$$

this minimum is shifted to finite values of β_ν and we have symmetry degeneracy with all its consequences.

It is interesting to note that the expression (8.64) contains a coupling of the phases of different clusters in its 4th order term. The phase coupling fixes only the sum $\varphi_{1\nu}+\varphi_{2\nu}$, whereas the phase difference $\varphi_{1\nu}-\varphi_{2\nu}$ is still arbitrary, as was found for a single signal-idler pair. Therefore, the signal phases and the idler phases separately are subject to a free diffusion.

The phase sum $\varphi_{1\nu}+\varphi_{2\nu}$ is obtained by minimizing ϕ^s. The phase sums of different clusters are therefore correlated. A calculation of the details of this phase coupling is usually difficult, since a strong coupling between the amplitudes and the phases has to be expected. The origin of this coupling can be seen by the following qualitative argument: The phases of signal and idler are determined, according to Eq. (8.64), by two main influences. These are the gain ($\sim F_p$), which has to be as large as possible, and the saturation ($\sim$ 4th order term), which has to be as small as possible, in order to minimize ϕ^s. The gain is a maximum if

$$\varphi_{1\nu}+\varphi_{2\nu}-\psi_\nu=0 \tag{8.67}$$

where ψ_ν is the phase of the complex coupling constant g_ν.

However, for this arrangement of phases the saturation is also a maximum, since now all terms in the double sum in Eq. (8.64) are positive and no compensation of different terms is possible. In order to have minimum saturation, the equidistribution of the quantity $\varphi_{1\nu}+\varphi_{2\nu}-\psi_\nu$ between 0 and 2π would be most favorable, since then the terms of the double sum could compensate each other. On the other hand, the amplitudes of all clusters for which the relation (8.67) does not hold, have to be small, since the gain is small. Therefore, a strong coupling between amplitudes and phases can be expected. In practical cases one usually has a small number of clusters. These are known to show an irregular spiking, which can be understood by the mechanism described above. Each cluster tries to fulfill the relation (8.67) itself, at the cost of the other clusters, which cannot fulfill relation (8.67) at the same time, because saturation would then be too strong. The stationary state is multistable, since each cluster may be the dominant one.

d) Multimode Laser Pump

All the calculations of this chapter can be generalized further by taking into account the multimode structure of the pumping laser. We simply have to replace the interaction Hamiltonian (8.3) by

$$H_{\text{int}}=\sum_\lambda H_\lambda \tag{8.68}$$

where each H_λ corresponds to one mode of the pumping laser and has the form given in Eq. (8.3)

$$H_\lambda = ih(F_{p\lambda} b_{p\lambda}^+ - F_{p\lambda}^* b_{p\lambda}) + ih\,\gamma_\lambda (b_{p\lambda} \Omega_\lambda^+ - b_{p\lambda}^+ \Omega_\lambda). \tag{8.69}$$

Here we have taken into account that the laser modes excite the medium directly at various frequencies $\omega_{p\lambda}$. The directly excited modes $b_{p\lambda}$, $b_{p\lambda}^+$ give rise to the same or to different scattering processes. These scattering processes are assumed to be of the same kind as in 8.1 and are described by the operators Ω_λ. All considerations of 8.1 remain valid, but they now refer only to one single mode of the pumping laser. We may use the new Hamiltonian to go through the analysis of 8.2, 8.3. Again we obtain a Fokker-Planck equation for the Wigner distribution of the process. Its stationary solution is given by $W_1^s \sim \exp(-\phi^s)$, with

$$\phi^s = 2\sum_\nu |\beta_\nu|^2 - 2\kappa^{-1} \sum_\lambda [\gamma_\lambda \kappa_{p\lambda}^{-1} (F_{p\lambda}^* \Omega_\lambda + F_{p\lambda} \Omega_\lambda^*) - \gamma_\lambda^2 \kappa_{p\lambda}^{-1} |\Omega_\lambda|^2]. \tag{8.70}$$

We realize, that W_1^s factorizes with respect to the contributions of the various laser modes. This factorizing does not exclude, however, that additional correlations between different resonant modes are produced by the multimode structure of the pump. In order to see this in more detail, and as an illustration of the general result (8.70), we investigate the stationary distribution for a second order parametric oscillator which is pumped by a multimode laser. The simplest situation occurs when each laser mode pumps its own signal-idler pair. In this case each function Ω_λ depends on its own pair of variables (amplitudes), and the distribution (8.59) factorizes with respect to different signal-idler pairs. No correlation between different signal-idler pairs occurs in this case. Another possibility is that all laser-modes excite the same dominating signal-mode β_0, β_0^*. However, because of the frequency differences between the laser modes, there corresponds a separate idler mode $\beta_\lambda, \beta_\lambda^*$ to each laser mode. In this case the potential ϕ^s takes the form

$$\begin{aligned}\phi^s = 2\left(|\beta_0|^2 + \sum_\lambda |\beta_\lambda|^2\right) - 2\kappa^{-1} \sum_\lambda \gamma_\lambda \kappa_{p\lambda}^{-1} (F_{p\lambda}^* \beta_0 \beta_\lambda + \text{c.c.}) \\ + 2\kappa^{-1} \sum_\lambda \gamma_\lambda^2 \kappa_{p\lambda}^{-1} |\beta_0 \beta_\lambda|^2 .\end{aligned} \tag{8.71}$$

The Wigner distribution $W_1^s \sim \exp(-\phi^s)$ factorizes with respect to the amplitudes of the various idler modes β_λ. Nevertheless, because they are all combined with the same signal mode β_0, there exists a strong correlation between the different idler modes, which will be considered below.

First we calculate the distribution of the signal mode amplitude by integrating over all complex idler amplitudes. We obtain

$$W_1^s(\beta_0, \beta_0^*) \sim \frac{1}{1+\sum_\lambda \gamma_\lambda^2(\kappa\kappa_{p\lambda})^{-1}|\beta_0|^2} \cdot \exp\left(-2|\beta_0|^2 + \frac{2\sum_\lambda \gamma_\lambda^2|F_{p\lambda}|^2(\kappa\kappa_{p\lambda})^{-2}|\beta_0|^2}{1+\sum_\lambda \gamma_\lambda^2(\kappa\kappa_{p\lambda})^{-1}|\beta_0|^2}\right). \tag{8.72}$$

For

$$\sum_\lambda \gamma_\lambda^2/\kappa\kappa_{p\lambda} \ll 1/|\beta_0|^2 \tag{8.73}$$

this may be reduced to

$$W_1^s \sim \exp\left[-2\left(1-\sum_\lambda \gamma_\lambda^2(\kappa\kappa_{p\lambda})^{-2}|F_{p\lambda}|^2\right)|\beta_0|^2 - 2\sum_{\lambda\lambda'} \gamma_\lambda^2\gamma_{\lambda'}^2|F_{p\lambda}|^2\kappa^{-3}\kappa_{p\lambda}^{-2}\kappa_{p\lambda'}^{-1}|\beta_0|^4\right]. \tag{8.74}$$

The distribution of the signal mode turns out to be of the same form as for single-mode pumping. The amplification due to the various pump modes is simply additive. The threshold condition for oscillation of this type is given by

$$\sum_\lambda \gamma_\lambda^2(\kappa\kappa_{p\lambda})^{-2}|F_{p\lambda}|^2 \geqq 1. \tag{8.75}$$

Because of the additivity of the gain due to the single laser modes, we obtain oscillation already for pump intensities $|F_{p\lambda}|^2$, which, in a single-mode pump, would not be sufficient to drive parametric oscillation. The saturation is described by the fourth order term of (8.74). This term contains a double sum over all modes and, hence, contains a factor proportional to the square of the number of pumping modes N_1. The maximum of the distribution (8.72) above threshold is at

$$|\beta_0^s|^2 = \frac{\sum_\lambda \gamma_\lambda^2\kappa_{p\lambda}^{-2}|F_{p\lambda}|^2 - \kappa^2}{2\sum_{\lambda\lambda'} \gamma_\lambda\gamma_{\lambda'}^2|F_{p\lambda}|^2\kappa^{-1}\kappa_{p\lambda}^{-2}\kappa_{p\lambda'}^{-1}}. \tag{8.76}$$

Because of the large saturation its order of magnitude is smaller by a factor N_1 if compared with single mode pumping with a pump intensity

$$|F_p|^2/\kappa_p^2 = \sum_\lambda |F_{p\lambda}|^2/\kappa_{p\lambda}^2 \tag{8.77}$$

and with

$$\gamma \approx \gamma_\lambda \,. \tag{8.78}$$

Finally, we consider the correlation between the different idler modes β_λ which is produced by their common signal mode. Integrating (8.71) over the complex β_0-plane we obtain

$$\begin{aligned} W_1^s(\{\beta,\beta^*\}) \sim & \frac{1}{1+\sum_\lambda \gamma_\lambda^2 \kappa^{-1} \kappa_{p\lambda}^{-1} |\beta_\lambda|^2} \\ & \cdot \exp\left[-2\sum_\lambda |\beta_\lambda|^2 + \frac{2\left|\sum_\lambda F_{p\lambda}^* \gamma_\lambda \kappa^{-1} \kappa_{p\lambda}^{-1} \beta_\lambda\right|^2}{1+\sum_\lambda \gamma_\lambda^2 \kappa^{-1}\kappa_{p\lambda}^{-1} |\beta_\lambda|^2}\right]. \end{aligned} \tag{8.79}$$

This expression no longer factorizes with respect to different idler modes. In particular (8.79) depends on the phases of the individual idler amplitudes and is invariant only against the rotation of the phase angle of the collective quantity $\sum_\lambda F_{p\lambda}^* \gamma_\lambda \kappa_{p\lambda}^{-1} \beta_\lambda$. This result has to be contrasted with the single mode case, where the idler phase was found to be arbitrary, if the signal phase was unknown. In the present case, the phase sum of the signal and each idler is determined by the phase of the corresponding mode of the pumping laser. Therefore, the correlation of the idler phases is nothing but an image of the correlation of the phases of the laser modes. If the pump light contains phase locked modes which give a periodic pulse train, then the idler phases are locked and give rise to a new pulse train. If the pump modes have random phases, the idler modes will also have random phases. The examples which we have considered in this section were only two possibilities out of a large manifold of mode configurations which may be realized in a given medium and a given cavity. Due to the lower threshold it seems possible to discriminate experimentally the case in which *one* signal mode is driven by *several* pump modes from other types of oscillation. In general, however, one has to expect that the formation of several clusters is more likely for multimode pumping and that experiments will not be reproducible. Therefore, there is at present no reason to give a further evaluation of our results for these cases.

Acknowledgement: It is a pleasure to thank Professor H. Haken for numerous discussions in which many of the viewpoints of this article have been developed for the first time. Thanks are further due to my friends K. Kaufmann, H. Geffers, and G. Krös who made the whole time during this work unforgettable. Finally, I want to thank the Physics Department of New York University, where the english version of this article was written, for its hospitality. In particular, I am indebted to Dr. W. Langer, who was so kind to help me with the language problems.

References

1. Glansdorff, P., Prigogine, I.: Physica **20**, 773 (1954); **30**, 351 (1964).
2. Prigogine, I., Glansdorff, P.: Physica **31**, 1242 (1965).
3. Glansdorff, P., Prigogine, I.: Physica **46**, 344 (1970).
4. Glansdorff, P., Prigogine, I.: Thermodynamic theory of structure, stability, and fluctuations. New York: Wiley Interscience 1971.
5. Schlögl, F.: Z. Physik **243**, 309 (1971); **244**, 199 (1971).
6. Chandrasekhar, S.: Hydrodynamic and hydromagnetic stability. Oxford: Clarendon Press 1961.
7. Prigogine, I., Nicolis, G.: J. Chem. Phys. **46**, 3542 (1967). — Prigogine, I., Lefever, R.: J. Chem. Phys. **48**, 1695 (1968).
8. see, e.g. Haken, H.: Encyclopedia of Physics, Vol. 25/2c. Berlin-Heidelberg-New York: Springer 1970.
9. see, e.g., Bloembergen, N.: Nonlinear Optics. New York: Benjamin 1965.
10. see, e.g., Quantum Optics, ed. Glauber, R. J. New York: Academic Press 1969. – Quantum Optics. Kay, S. M., Maitland, A. (Eds.). New York: Academic Press 1970.
11. see, e.g., Yariv, A., Louisell, W. H.: IEEE J. Quantum Electronics **2**, 418 (1966). – Huth, B. G., Karlov, N. V., Pantell, R. H., Puthoff, H. E.: IEEE J. Quant. Electr. **2**, 763 (1966).
12. Hohenberg, P. C.: Phys. Rev. **158**, 383 (1967). – Mermin, N. D., Wagner, H.: Phys. Rev. Letters **17**, 1133 (1966). – Mermin, N. D.: J. Math. Phys. **8**, 1061 (1967).
13. see, e.g., De Groot, S. R., Mazur, P.: Nonequilibrium thermodynamics. Amsterdam: North Holland 1962.
14. Onsager, L., Machlup, S.: Phys. Rev. **91**, 1505 (1953); **91**, 1512 (1953).
15. Hashitsume, N.: Progr. Theor. Phys. **8**, 461 (1952); **15**, 369 (1956).
16. Tisza, L., Manning, I.: Phys. Rev. **105**, 1695 (1957).
17. Landau, L. D., Lifshitz, E. M.: Course of Theoretical Physics, Vol. 5. London: Pergamon 1958.
18. Armstrong, J. A., Smith, A. W.: Progress in Optics, Vol. 6, ed. Wolf, E. Amsterdam: North Holland 1967. – Arecchi, F. T., Rodari, G. S., Sona, A.: Phys. Letters **25**A, 59 (1967).
19. Risken, H.: Progress in Optics, Vol. 8, ed. Wolf, E. Amsterdam: North Holland 1970.
20. Stratonovich, R. L.: Topics in the Theory of Random Noise, Vol. 1. New York: Gordon and Breach 1963.
21. Lax, M.: Rev. Mod. Phys. **32**, 25 (1960); **38**, 359 (1966); **38**, 541 (1966).
22. Feynman, R. P., Hibbs, A. R.: Quantum mechanics and path integrals. New York: McGraw Hill 1965.
23. Statistical Mechanics, Rice, S. A., Freed, K. F., Light, J. C. (Eds.). Chicago: University of Chicago Press 1972.
24. see, e.g., Hahn, W.: Stability of motion. Berlin-Heidelberg-New York: Springer 1967.
25. Lebowitz, J. L., Bergmann, P. G.: Ann. Phys. **1**, 1 (1957).
26. Weidlich, W.: Z. Physik **248**, 234 (1971).
27. Agarwal, G. S.: Z. Physik **252**, 25 (1972).
28. Landsberg, P. T.: Thermodynamics. New York: Interscience 1961.
29. Graham, R., Haken, H.: Z. Physik **243**, 289 (1971).
30. Graham, R., Haken, H.: Z. Physik **245**, 141 (1971).
31. Klein, M. J.: Phys. Rev. **97**, 1446 (1955).
32. Haken, H.: Z. Physik **219**, 246 (1969).
33. Green, M. S.: J. Chem. Phys. **20**, 1281 (1952).
34. Zwanzig, R.: Phys. Rev. **124**, 983 (1961).
35. Onsager, L.: Phys. Rev. **37**, 405 (1931); **38**, 2265 (1931).

36. Casimir, H. B. G.: Rev. Mod. Phys. **17**, 343 (1945).
37. Bjorkholm, J. E.: Appl. Phys. Letters **13**, 53 (1968). – Kreuzer, L. B.: Appl. Phys. Letters **13**, 57 (1968). – Falk, J., Murray, J. E.: Appl. Phys. Letters **14**, 245 (1969).
38. Haake, F.: Z. Physik **227**, 179 (1969).
39. Arecci, F. T., Corti, M., De Giorgio, V., Vendramini, A.: paper presented at the 6th IQEC, Kyoto, 1970.
40. Mandel, L.: Proc. Phys. Soc. **72**, 1037 (1958); Progress in Optics, Vol. 2, ed. Wolf, E. Amsterdam: North Holland 1963.
41. Hanbury-Brown, R., Twiss, R. O.: Nature **177**, 27 (1956).
42. Ernst, V., Stehle, P.: Phys. Rev. **176**, 1456 (1968).
43. see, e.g., Argyres, P. N., Kelley, P. L.: Phys. Rev. **134**, A 98 (1964). – Haake, F.: Z. Physik **223**, 353, 364 (1969).
44. Bloembergen, N., Shen, Y. R.: Phys. Rev. **133**, A 37 (1964).
45. Graham, R., Haken, H.: Z. Physik **213**, 420 (1968); **237**, 31 (1970).
46. Scully, M., De Giorgio, V.: Phys. Rev. **A 2**, 1170 (1970).
47. Thomas, H. in Structural Phase Transitions and Soft Modes, Samuelsen, E. J., Andersen, E., Feder, J. (Eds.), Oslo: Universitätsforlaget 1971.
48. see, e.g., Van Hove, L.: Phys. Rev. **95**, 1374 (1954). – Landau, L. D., Khalatnikov, I. M.: Dokl. Akad. Nauk SSSR **90**, 469 (1954).
49. Korenmann, V.: Ann. Phys. **39**, 72 (1966).
50. Haken, H.: Z. Physik **181**, 96 (1964).
51. Risken, H.: Z. Physik **186**, 85 (1965).
52. see, e.g., Armstrong, J. A., Smith, A. W.: Progress in Optics, Vol. 6, ed. Wolf. Amsterdam: North Holland 1967.
53. Risken, H.: Z. Physik **191**, 302 (1966). – Risken, H., Vollmer, H. D.: Z. Physik **201**, 323 (1967); **204**, 240 (1967). – Risken, H.: Fortschr. Phys. **16**, 261 (1968). – Hempstead, R. D., Lax, M.: Phys. Rev. **161**, 350 (1967).
54. Grossmann, S., Richter, P. H.: Z. Physik **242**, 458 (1971).
55. Richter, P. H., Grossmann, S.: Z. Physik **248**, 244 (1971); **255**, 59 (1972).
56. Grossmann, S., Richter, P. H.: Z. Physik **249**, 43 (1971).
57. Haken, H.: Z. Physik **219**, 246 (1969).
58. Haken, H., Sauermann, H.: Z. Physik **173**, 261 (1963); **176**, 47 (1963).
59. Landau, L. D., Lifshitz, E. M.: Course of Theoretical Physics, Vol. 6. London: Pergamon 1959.
60. Lethokov: Zh. Eksp. Teoret. Fiz. **55**, 1077 (1968).
61. see, e.g., Pelikan, H.: Z. Physik **201**, 523 (1967); **211**, 418 (1968).
62. Kaufmann, K., Weidlich, W.: Z. Physik **217**, 113 (1968).
63. Statz, H., Tang, C. L.: J. Appl. Phys. **36**, 3963 (1965). – Statz, H., De Mars, G. A., Tang, C. L.: J. Appl. Phys. **38**, 2212 (1967). – Tang, C. L., Statz, H.: J. Appl. Phys. **38**, 2963 (1967).
64. Haken, H., Sauermann, H., Schmidt, Ch., Vollmer, H. D.: Z. Physik **206**, 369 (1967).
65. Haken, H., Panthier, M.: IEEE J. Quant. Electr. **1**, 12 (1965).
66. Hübner, H.: Z. Physik **239**, 103 (1970).
67. Landau, L. P., Ginzburg, V. L.: JETP **20**, 1064 (1950); see also Schmidt, H.: Z. Physik **216**, 336 (1968).
68. Kac, M.: Probability and related topics in physical sciences, Chap. 4. New York: Interscience 1959.
69. Little, W. A.: Phys. Rev. **156**, 396 (1967).
70. Risken, H., Nummedal, K.: Phys. Letters **26 A**, 275 (1968); J. Appl. Phys. **39**, 4662 (1968).
71. Landau, L. D.: JETP **7**, 627 (1937).
72. Louisell, W. H.: Coupled mode and parametric electronics. New York: John Wiley 1960; Kleinmann, D. A.: Phys. Rev. **128**, 1761 (1962).

73. Kirsanov, B.P. Nonlinear optics. Proc. P. N. Lebedev Phys. Inst. **43**, 171 (1970).
74. Louisell, W.H., Yariv, A., Siegmann, A.E.: Phys. Rev. **124**, 1646 (1961). – Gordon, J.P., Louisell, W.H., Walker, L.R.: Phys. Rev. **129**, 481 (1963). – Louisell, W.H.: Radiation and noise in quantum electronics. New York: McGraw Hill 1964. Wagner, W.G., Hellwarth, R.W.: Phys. Rev. **133**, A 915 (1964).
75. Mollow, B.R., Glauber, R.J.: Phys. Rev. **160**, 1076 (1967); **160**, 1087 (1967).
76. Graham, R., Haken, H.: Z. Physik **210**, 276 (1968).
77. Graham, R.: Z. Physik **210**, 319 (1968).
78. Graham, R.: Z. Physik **211**, 469 (1968). – White, D.R., Louisell, W.H.: Phys. Rev. **A1**, 1347, (1970).
79. Graham, R.: p. 489 in Quantum Optics, ed. Kay, S.M., Maitland, A. New York: Academic Press 1970.
80. Graham, R.: Phys. Letters **32A**, 373 (1970).
81. Giordmaine, J.A., Miller, R.C.: p. 31 in Physics of Quantum Electronics, ed. Kelley, P.L., Lax, B., Tannenwald, P.E. New York: McGraw Hill 1966.
82. Landauer, R., Woo, J.: IEEE J. Quant. Electr. **7**, 435 (1971).
83. Risken, H., Schmid, Ch., Weidlich, W.: Z. Physik **193**, 37 (1966).
84. Glauber, R.J.: Lectures on optical coherence and photon statistics. New York: Gordon and Breach 1964.
85. Glauber, R.J.: Phys. Rev. **130**, 2529 (1963); **131**, 2766 (1963). – Sudarshan, E.C.C.: Phys. Rev. Lett. **10**, 277 (1963).
86. Lax, M., Louisell, W.H.: IEEE J. Quant. Electr. **3**, 47 (1967).
87. Wigner, F.P.: Phys. Rev. **40**, 749 (1932). – Moyal, J.E.: Proc. Cambridge Phil. Soc. **45**, 99 (1949).
88. Graham, R., Haake, F., Haken, H., Weidlich, W.: Z. Physik **213**, 21 (1967).
89. Haken, H.: Opt. Electronics **2**, 161 (1970).

Dr. R. Graham
New York University
Department of Physics
New York, N. Y. 10003, USA

on leave of absence from:

Institut für Theoretische Physik
der Universität Stuttgart
D-7000 Stuttgart 1
Federal Republic of Germany

Statistical Treatment of Open Systems by Generalized Master Equations

FRITZ HAAKE

Contents

1. Introduction

This paper deals with the dynamics of open systems ($\mathfrak{S}$) moving irreversibly under the influence of their surroundings ($\mathfrak{B}$). As a basis for the discussion of an open system $\mathfrak{S}$ we use a complete microscopic description of the composite system $\mathfrak{S} \oplus \mathfrak{B}$. By eliminating the coordinates of $\mathfrak{B}$ we infer the behavior of $\mathfrak{S}$. The motivation for this investigation is that nature frequently confronts us with coupled systems $\mathfrak{S}$ and $\mathfrak{B}$

only one of which, say $\mathfrak{S}$, is of experimental relevance. It is then a dictate of economy to look for a "closed" description of the dynamics of $\mathfrak{S}$ alone. Let us mention just three out of the countless examples.

(1) The motion of penduli is usually found empirically to be describable in terms of the equation of motion of an ideal oscillator augmented by a suitable friction term. Statistical mechanics explains this behavior by accounting for the coupling of the pendulum ($\mathfrak{S}$) to its surroundings ($\mathfrak{B}$). When the coordinates of $\mathfrak{B}$ are eliminated from the equations of motion for all degrees of freedom of $\mathfrak{S} \oplus \mathfrak{B}$ the influence of $\mathfrak{B}$ on $\mathfrak{S}$ is found, under certain conditions, to amount to a friction force.

(2) In light scattering experiments on simple liquids one usually observes long-wavelength transport processes like heat diffusion and sound waves. There is a macroscopic theory of the long-wavelength behaviour of liquids, namely hydrodynamics. The set of hydrodynamic variables (number density of molecules, energy density, velocity of molecules averaged over volume elements large compared to intermolecular distances, etc.) may be looked upon as an open system $\mathfrak{S}$ with all other degrees of freedom of the liquid constituting a "surrounding" $\mathfrak{B}$. The statistical-mechanical derivation of hydrodynamics requires the elimination of the coordinates of $\mathfrak{B}$ from the microscopically complete description of the liquid $\mathfrak{S} \oplus \mathfrak{B}$.

(3) Experiments on lasers refer to the radiation output ($\mathfrak{S}$) and never to the active atoms nor the various pump and loss mechanisms ($\mathfrak{B}$) involved. While the theory has to be based upon a description of all the interacting components of the laser system, it is natural that it should aim as it has at setting up dynamic equations for the experimentally relevant radiation field ($\mathfrak{S}$) alone.

The physical systems treated in some detail in the present paper are the damped harmonic oscillator, superconductors, superradiant devices, the laser, and the Heisenberg magnet near the Curie Temperature. In these cases we have as the respective open system $\mathfrak{S}$ and its surrounding $\mathfrak{B}$: the ideal oscillator and a heat bath, electrons ($\mathfrak{S}$) and phonons ($\mathfrak{B}$) for a superconductor, radiating atoms ($\mathfrak{S}$) and radiated light ($\mathfrak{B}$) in the case of superradiance, radiation field ($\mathfrak{S}$) and active atoms as well as pump and loss mechanisms ($\mathfrak{B}$) for the laser, and, finally, long-wavelength ($\mathfrak{S}$) and short-wavelength ($\mathfrak{B}$) spin fluctuations for the magnet.

Since the systems mentioned physically have scarcely anything in common (beyond being of the structure $\mathfrak{S} \oplus \mathfrak{B}$), it is not surprising that a vast variety of different formal techniques have been used in the literature in dealing with them. Rather than giving a survey of various formalisms we here present a unified treatment of open systems in terms of generalized master equations.

Master equations were first introduced into quantum statistical mechanics by Pauli [1] to describe the relaxation of macroscopic systems into thermal equilibrium. In their original form used by Pauli they are rate equations for occupation numbers of quantum levels which are dynamically connected by suitably chosen transition rates. Pauli's derivation of master equations from Schrödinger's equation was based upon the assumption that the expansion coefficients of the wave function in an expansion in terms of energy-eigenfunctions have random phases at all times. This assumption makes possible a dynamical description of the system in terms of occupation numbers of energy levels rather than in terms of the complex probability amplitudes of the wave function with respect to energy eigenstates. Later work by van Hove [2], Nakajima [3], Zwanzig [4], Montroll [5], and Prigogine and Resibois [6] has shown that the unsatisfactory assumption of continuously random phases is unnecessary. Pauli's master equations have proved to be special cases of rigorous socalled generalized master equations. For a survey of these modern master equations we refer the reader to [7]. In the present paper we will use a generalized master equation constructed independently by Nakajima [3] and Zwanzig [4].

The Nakajima-Zwanzig theory will be discussed in detail in Section II. Let us at this point make just a few qualitative introductory remarks on it. The starting point is the wellknown equation of motion for the density operator of the composite system $\mathfrak{S}\oplus\mathfrak{B}$

$$\dot{W}(t) = -(i/\hbar)\,[H, W(t)] \equiv -i L W(t) \tag{1.1}$$

where $W(t)$ and H denote the density operator and the Hamiltonian, respectively. Only that part of the information contained in the Liouville-von Neumann equation (1.1) which refers to the subsystem $\mathfrak{S}$ is considered relevant. By using a certain projection operator $\mathfrak{P}(\mathfrak{P}^2 = \mathfrak{P})$ a reduced density operator $\varrho(t)$ for the open system $\mathfrak{S}$ is obtained from the full density operator $W(t)$. Schematically,

$$W(t) \xrightarrow{\mathfrak{P}} \varrho(t)\,, \tag{1.2}$$

which leaves open, for the moment being, how the full density operator $W(t)$ is to be operated upon by the projector $\mathfrak{P}$ to yield the reduced density operator $\varrho(t)$. With a suitable definition of what (1.2) precisely means one finds that the Liouville-von Neumann equation (1.1) entails an equation of motion for the reduced density operator $\varrho(t)$ of $\mathfrak{S}$ which is of the following form

$$\dot{\varrho}(t) = -i L_{\text{eff}}\,\varrho(t) + \int_0^t \mathrm{d}t'\, K(t,t')\,\varrho(t') + I(t)\,. \tag{1.3}$$

Expressions for the effective Liouvillian L_{eff}, the integral kernel $K(t, t')$, and the inhomogeneity $I(t)$ will be given in Section II.

As an illustration of the potential usefulness of the theory behind Eq. (1.3) let us briefly mention two different applications. The first is the one Nakajima and Zwanzig had in mind when constructing the general theory. It is concerned with the above-mentioned problem posed by Pauli: how do the occupation probabilities for the energy levels of a macroscopic system relax to an equilibrium distribution, starting out from some arbitrary initial distribution. To attack this problem Nakajima and Zwanzig consider a Hamiltonian $H = H_0 + H_1$ consisting of a main part H_0 and a small perturbation H_1. The set of all diagonal elements $\langle n|W(t)|n\rangle$ of the density operator with respect to the eigenstates of $H_0 (H_0|n\rangle = E_n|n\rangle)$ is taken as the open system $\mathfrak{S}$ which interacts with the surrounding $\mathfrak{B}$ constituted by the off-diagonal matrix elements of $W(t)$. The projector $\mathfrak{P}$ is then chosen as $\mathfrak{P}|n\rangle \langle n'| = \delta_{nn'} |n\rangle \langle n|$ whereupon the reduced density operator of $\mathfrak{S}$, defined as $\varrho(t) = \mathfrak{P} W(t)$, becomes just the diagonal part of $W(t)$ in the H_0-representation. The procedure leading from the Liouville-von Neumann equation (1.1) to the generalized master equation (1.3) then amounts to eliminating the off-diagonal part of $W(t)$ from Eq. (1.1). Under appropriate conditions for the Hamiltonians H_0 and H_1 and for the initial state $W(0)$ Eq. (1.3) can be shown to reduce to Pauli's original master equation [8]. – The second application we want to mention here has first been made by Argyres and Kelley [9]. These authors consider a spin system ($\mathfrak{S}$) weakly coupled to some large system in thermal equilibrium ($\mathfrak{B}$), i.e. a heat bath. The reduced density operator $\varrho(t)$ of the spin system $\mathfrak{S}$ is defined as the partial trace of the full density operator $W(t)$ of $\mathfrak{S} \oplus \mathfrak{B}$, $\varrho(t) = \text{tr}_{\mathfrak{B}} W(t)$. Under suitable conditions for the coupling of the spin system $\mathfrak{S}$ to the heat bath $\mathfrak{B}$ the generalized master equation then describes the relaxation of the spins into thermal equilibrium. – We have intentionally mentioned these two applications here, partly because of their historical importance but mainly because they are so different physically. They indicate the remarkable flexibility of the Nakajima-Zwanzig theory.

In spite of the flexibility of the generalized master equation (1.3) stressed above there are limits to its practical usefulness which it is appropriate to underscore here as well. If Eq. (1.3) is to be used in describing the motion of a given open system, the rather involved formal expressions for the integral kernel $K(t)$ and the inhomogeneity $I(t)$ have to be evaluated explicitly first. The evaluation of K and I generally requires series expansions of these quantities. If for a given problem there are no small dimensionless parameters in terms of which such expansions can be generated, the generalized master equation (1.3) remains an empty concept. We will therefore put special emphasis, in all the applications to be presented in this paper, on identifying the respective relevant small parameters.

The paper is organized as follows: Section II gives a detailed discussion of the projector technique. In Sections III through VII we discuss the applications to linear damping phenomena, superconductors, superradiance, lasers, and the critical dynamics of the Heisenberg magnet. Each section is headed by a separate introduction.

2. Generalized Master Equations

2a) Introductory Remarks

We here derive and discuss various modifications of the generalized master equation (1.3), using Zwanzig's projector technique. The first case, considered in subsection (2b) is that of a closed system $\mathfrak{S} \oplus \mathfrak{B}$ whose Hilbertspace $\mathfrak{H}_{\mathfrak{S}\oplus\mathfrak{B}}$ has the property $\mathfrak{H}_{\mathfrak{S}\oplus\mathfrak{B}} = \mathfrak{H}_{\mathfrak{S}} \otimes \mathfrak{H}_{\mathfrak{B}}$. The reduced density operator of the open system $\mathfrak{S}$ is $\varrho(t) = \mathrm{tr}_{\mathfrak{B}} W(t)$. Next, in subsection (2c) we drop the condition that $\mathfrak{S} \oplus \mathfrak{B}$ be closed. If $\mathfrak{S} \oplus \mathfrak{B}$ is an open system itself since there are time-dependent external fields acting on it, the Hamiltonian $H(t)$ and the Liouvillian $L(t)$ display an explicit time dependence; consequently, operator products occuring in the generalized master equation for $\varrho(t)$ of $\mathfrak{S}$ have to be time-ordered. On the other hand, $\mathfrak{S} \oplus \mathfrak{B}$ may be an open system because it moves irreversibly under the influence of some other system $\mathfrak{C}$; in such a case the equation of motion for $W(t)$ of $\mathfrak{S} \oplus \mathfrak{B}$ may still be, in simple cases, of the structure $\dot{W} = -iLW$; however, the Liouvillian is no longer defined as the commutator with a Hamiltonian; yet a generalized master equation still governs the behaviour of ϱ of $\mathfrak{S}$. Then, in subsection (2d) we deal with a situation where the dynamics of $\mathfrak{S} \oplus \mathfrak{B}$ is described by a reversible or irreversible equation of motion for a quasiprobability distribution W over c-number variables; this equation of motion is assumed to be a first-order differential equation in time, i.e. to have the form $\dot{W} = -iLW$; the set of c-number variables is separated in two subsets, $\mathfrak{S}$ and $\mathfrak{B}$; the generalized master equation for a reduced quasi-probability distribution ϱ over the set of variables $\mathfrak{S}$ is derived; this is the most general version of the Nakajima-Zwanzig equation, since it holds regardless of the structure of the Hilbertspace $\mathfrak{H}_{\mathfrak{S}\oplus\mathfrak{B}}$ and of whether the motion of $\mathfrak{S} \oplus \mathfrak{B}$ is reversible or irreversible. Finally, in subsection (2.e), we use a formal integral of the generalized master equation (1.3) to construct expressions for multi-time correlation functions of observables of the open system $\mathfrak{S}$; these expressions can be evaluated once the solution of Eq. (1.3) is known; they can also be used to generate hierarchies of equations of motion for correlation functions or Green's functions which it may be easier to solve than the generalized master equation (1.3) itself.

Each of these modifications of the original Nakajima-Zwanzig equation will be needed in one or several of the applications given later in Section III–VII.

2b) Open System $\mathfrak{S}$ as Subsystem of a Closed System $\mathfrak{S} \oplus \mathfrak{B}$

We consider a closed system $\mathfrak{S} \oplus \mathfrak{B}$ composed of two interacting parts, $\mathfrak{S}$ and $\mathfrak{B}$. The observables of $\mathfrak{S}(\mathfrak{B})$ are represented by operators $S_1, S_2, \ldots (B_1, B_2, \ldots)$ in a Hilbertspace $\mathfrak{H}_{\mathfrak{S}}(\mathfrak{H}_{\mathfrak{B}})$. The Hilbertspace $\mathfrak{H}_{\mathfrak{S} \oplus \mathfrak{B}}$ of the composite system is the direct product $\mathfrak{H}_{\mathfrak{S} \oplus \mathfrak{B}} = \mathfrak{H}_{\mathfrak{S}} \otimes \mathfrak{H}_{\mathfrak{B}}$. Physically, this means that $\mathfrak{S}$ and $\mathfrak{B}$ become physical systems, each in its own right, if their interaction is switched off. The state of $\mathfrak{S} \oplus \mathfrak{B}$ is described by the density operator $W(t)$ which obeys the Liouville-von Neumann equation

$$\dot{W}(t) = -(i/\hbar)\,[H, W(t)] = -iLW(t)\,. \tag{2b.1}$$

The Hamiltonian H and, correspondingly, the Liouvillian L consist of three parts

$$\begin{aligned} H &= H_{\mathfrak{S}} + H_{\mathfrak{B}} + H_{\mathfrak{S}\mathfrak{B}} \\ L &= L_{\mathfrak{S}} + L_{\mathfrak{B}} + L_{\mathfrak{S}\mathfrak{B}} \end{aligned} \tag{2b.2}$$

referring to the free motion of $\mathfrak{S}$ and $\mathfrak{B}$ and an interaction, respectively. If the solution $W(t)$ of Eq. (2b.1) is known, expectation values of observables of $\mathfrak{S} \oplus \mathfrak{B}$ may be evaluated as, e.g.,

$$\langle B_i S_j \rangle = \operatorname{tr} B_i S_j W. \tag{2b.3}$$

We now assume that only $\mathfrak{S}$ is experimentally relevant, i.e. that the interesting expectation values are

$$\langle S_1 S_2 \ldots S_n \rangle = \operatorname{tr} S_1 S_2 \ldots S_n W(t)\,. \tag{2b.4}$$

Since the trace operation can be carried out in two steps,

$$\operatorname{tr} = \operatorname{tr}_{\mathfrak{S}} \operatorname{tr}_{\mathfrak{B}}\,, \tag{2b.5}$$

the expectation values of observables of $\mathfrak{S}$ can be written in terms of the reduced density operator of $\mathfrak{S}$

$$\varrho(t) = \operatorname{tr}_{\mathfrak{B}} W(t) \tag{2b.6}$$

as

$$\langle S_1 S_2 \ldots S_n \rangle = \operatorname{tr}_{\mathfrak{S}} S_1 S_2 \ldots S_n \varrho(t)\,. \tag{2b.7}$$

This suggests to look for a "closed" description of the dynamics of $\mathfrak{S}$, that is to construct an equation of motion for the experimentally relevant reduced density operator $\varrho(t)$. Such an equation should follow from the Liouville-von Neumann equation (2b.1) by eliminating from it the coordinates of the irrelevant subsystem $\mathfrak{B}$. In fact, such an elimination can be carried out formally using a method designed by Nakjima [3] and Zwanzig [4]. Let us briefly sketch this procedure. The full density operator $W(t)$ is decomposed into two parts using a projector

$$W(t) = \mathfrak{P}\, W(t) + (1-\mathfrak{P})\, W(t)\,, \quad \mathfrak{P}^2 = \mathfrak{P}\,. \tag{2b.8}$$

The projector $\mathfrak{P}$ is defined as

$$\mathfrak{P} = B_{\text{ref}}\, \text{tr}_{\mathfrak{B}}\,, \quad \text{tr}_{\mathfrak{B}} B_{\text{ref}} = 1\,. \tag{2b.9}$$

As a consequence of this definition

$$\mathfrak{P}\, W(t) = B_{\text{ref}}\, \varrho(t) \tag{2b.10}$$

is the relevant part of $W(t)$ containing all information with respect to the subsystem $\mathfrak{S}$ whereas the irrelevant part $(1-\mathfrak{P})\, W(t)$ takes up the information with respect to the subsystem $\mathfrak{B}$ and to correlations between $\mathfrak{S}$ and $\mathfrak{B}$. The parameter B_{ref} occuring in $\mathfrak{P}$ may be chosen arbitrarily within the indicated constraint of normalization, $\text{tr}_{\mathfrak{B}} B_{\text{ref}} = 1$. As we will see below it plays the role of a reference state of the system $\mathfrak{B}$ to be eliminated. The physically important question of how to best choose this reference state will also be discussed at the end of this subsection. The formal operations to be presented now are independent of how B_{ref} is chosen. By inserting the decomposition (2b.8) in Eq. (2b.1) and acting on this equation from the left with $\mathfrak{P}$ and $(1-\mathfrak{P})$, respectively, we get two coupled equations for $\mathfrak{P}\, W$ and $(1-\mathfrak{P})\, W$

$$\begin{aligned} \mathfrak{P}\, \dot{W}(t) &= -i\mathfrak{P} L \mathfrak{P}\, W(t) - i\mathfrak{P} L (1-\mathfrak{P})\, W(t) \\ (1-\mathfrak{P})\, \dot{W}(t) &= -i(1-\mathfrak{P})\, L\mathfrak{P}\, W(t) - i(1-\mathfrak{P})\, L(1-\mathfrak{P})\, W(t)\,. \end{aligned} \tag{2b.11}$$

The second of these equations can be integrated formally to yield the irrelevant part of $W(t)$, $(1-\mathfrak{P})\, W(t)$, in terms of the relevant part $\mathfrak{P}\, W(t)$

$$\begin{aligned} (1-\mathfrak{P})\, W(t) &= \exp[-i(1-\mathfrak{P})\, Lt]\,(1-\mathfrak{P})\, W(0) \\ &\quad - i\int_0^t dt' \exp[-i(1-\mathfrak{P}) Lt']\,(1-\mathfrak{P}) L \mathfrak{P}\, W(t-t')\,. \end{aligned} \tag{2b.12}$$

A closed equation of motion for the relevant part, $\mathfrak{P}\, W(t)$, is then obtained by inserting the formal integral (2b.12) into the first of Eqs. (2b.11). This gives, after performing the partial trace $\text{tr}_{\mathfrak{B}}$ the generalized master equation for the reduced density operator $\varrho(t)$ of $\mathfrak{S}$

$$\dot{\varrho}(t) = -i L_{\text{eff}}\, \varrho(t) + \int_0^t dt'\, K(t')\, \varrho(t-t') + I(t)\,. \tag{2b.13}$$

The effective Liouvillian L_{eff}, the integral kernel $K(t)$, and the inhomogeneity $I(t)$ come out as

$$\begin{aligned}
L_{\text{eff}} &= \text{tr}_{\mathfrak{B}}\, L B_{\text{ref}} = L_{\mathfrak{S}} + \text{tr}_{\mathfrak{B}}\, L_{\mathfrak{S}\mathfrak{B}}\, B_{\text{ref}} \\
K(t) &= -\text{tr}_{\mathfrak{B}}\, L \exp[-i(1-\mathfrak{P})\, Lt]\, (1-\mathfrak{P})\, L B_{\text{ref}} \\
&= -\text{tr}_{\mathfrak{B}}\, L_{\mathfrak{S}\mathfrak{B}} \exp[-i(1-\mathfrak{P})\, Lt]\, (1-\mathfrak{P})\, (L_{\mathfrak{B}} + L_{\mathfrak{S}\mathfrak{B}})\, B_{\text{ref}} \\
I(t) &= -i\, \text{tr}_{\mathfrak{B}}\, L \exp[-i(1-\mathfrak{P})\, Lt]\, (1-\mathfrak{P})\, W(t=0) \\
&= -i\, \text{tr}_{\mathfrak{B}}\, L_{\mathfrak{S}\mathfrak{B}} \exp[-i(1-\mathfrak{P})\, Lt]\, (1-\mathfrak{P})\, W(0)\,.
\end{aligned} \tag{2b.14}$$

In simplifying these expressions we have used the decomposition (2b.2) for the Liouvillian L and the identities

$$\begin{aligned}
\text{tr}_{\mathfrak{B}}\, L_{\mathfrak{B}} &= 0 \\
\text{tr}_{\mathfrak{B}}\, L_{\mathfrak{S}} &= L_{\mathfrak{S}}\, \text{tr}_{\mathfrak{B}}
\end{aligned} \tag{2b.15}$$

following from the cyclic invariance of the trace and the commutativity of operations in $\mathfrak{H}_{\mathfrak{S}}$ and $\mathfrak{H}_{\mathfrak{B}}$, respectively.

The generalized master equation is an inhomogeneous integro-differential equation in time. It describes how the open system $\mathfrak{S}$ moves under the influence of $\mathfrak{B}$. It is formally exact. Together with the relation (2b.12) it is equivalent to the Liouville-von Neumann equation (2b.1). In the trivial case of no interaction between $\mathfrak{S}$ and $\mathfrak{B}$ it reduces, of course, to the Liouville-von Neumann equation $\dot{\varrho} = -i L_{\mathfrak{S}}\, \varrho$ for the then closed system $\mathfrak{S}$.

For later use we want to provide ourselves with a formal integral of the Nakajima-Zwanzig equation (2b.13)

$$\varrho(t) = V(t)\, \varrho(0) + \int_0^t dt'\, V(t')\, I(t-t')\,. \tag{2b.16}$$

This defines the integral operator $V(t)$ such that $V(t)\,\varrho(0)$ solves the homogeneous part of Eq. (2b.13). We may look upon $V(t)$ as a time evolution operator for $\mathfrak{S}$ since it uniquely relates the density operator $\varrho(t)$ with its initial value $\varrho(0)$ if the inhomogeneity $I(t)$ is known. The definition of $V(t)$ implies the following formal properties

$$\begin{aligned}
\dot{V}(t) &= -i L_{\text{eff}}\, V(t) + \int_0^t dt'\, K(t')\, V(t-t') \\
V(0) &= 1 \\
\text{tr}_{\mathfrak{S}}\, V(0) &= \text{tr}_{\mathfrak{S}}\,.
\end{aligned} \tag{2b.17}$$

Let us now discuss the conditions under which the Nakajima-Zwanzig theory sketched above can be put to practical use. A necessary condition certainly is that it must be possible, for a given physical system, to explicitly evaluate the rather involved formal expressions (2b.14) for $K(t)$ and $I(t)$. Such an evaluation in general requires a perturbation expansion of the exponential $\exp[-i(1-\mathfrak{P})(L_{\mathfrak{S}}+L_{\mathfrak{B}}+L_{\mathfrak{S}\mathfrak{B}})t]$ occurring in (2b.14) in terms of the interaction Liouvillian $L_{\mathfrak{S}\mathfrak{B}}$. The resulting perturbation series for $K(t)$ and $I(t)$ are easily written down formally and read, for the Laplace transforms of these quantities,

$$\begin{aligned} K(z) &= \int_0^\infty dt\, e^{-zt} K(t) \\ &= -\sum_{n=0}^{\infty} \mathrm{tr}_{\mathfrak{B}}\, L_{\mathfrak{S}\mathfrak{B}}\, U(z)\, [-i(1-\mathfrak{P})\, L_{\mathfrak{S}\mathfrak{B}}\, U(z)]^n \\ &\quad \cdot (1-\mathfrak{P})\, (L_{\mathfrak{B}}+L_{\mathfrak{S}\mathfrak{B}})\, B_{\mathrm{ref}} \\ I(z) &= \int_0^\infty dt\, e^{-zt} I(t) \\ &= -i \sum_{n=0}^{\infty} \mathrm{tr}_{\mathfrak{B}}\, L_{\mathfrak{S}\mathfrak{B}}\, U(z)\, [-i(1-\mathfrak{P})\, L_{\mathfrak{S}\mathfrak{B}}\, U(z)]^n (1-\mathfrak{P})\, W(0) \end{aligned} \tag{2b.18}$$

with

$$U(z) = \int_0^\infty dt\, e^{-zt} \exp[-i(L_{\mathfrak{S}}+L_{\mathfrak{B}})\,t] = [z+i(L_{\mathfrak{S}}+L_{\mathfrak{B}})]^{-1}\,.$$

However, these perturbation series make sense only if they can be identified to go in terms of a small dimensionless parameter which is, formally, $O(L_{\mathfrak{S}\mathfrak{B}})/O(L_{\mathfrak{S}}+L_{\mathfrak{B}})$. If the interaction between $\mathfrak{S}$ and $\mathfrak{B}$ is too strong for the expansions (2b.18) to converge, the Nakajima-Zwanzig equation is in general useless. If, on the other hand, the series do go in terms of a small dimensionless parameter, the use of the generalized master equation offers substantial advantages over other methods of evaluating $\varrho(t)$ as, e.g., the perturbation theoretical solution of Eq. (2b.1). This latter method proceeds by expanding the time evolution operator in $\varrho(t)=\mathrm{tr}_{\mathfrak{B}}\exp[-i(L_{\mathfrak{S}}+L_{\mathfrak{B}}+L_{\mathfrak{S}\mathfrak{B}})t]\cdot W(0)$ in terms of $L_{\mathfrak{S}\mathfrak{B}}$. It is easy to see that an infinite number of terms of all orders of this elementary perturbation expansion have to be summed up in order to recover a given finite order approximation to the series (2b.18).

Another point we want to discuss here concerns the physical meaning of the parameter B_{ref} occurring in the definition (2b.9) of the projector $\mathfrak{P}$ and the question of how to best choose it. Let us emphasize again that the Nakajima-Zwanzig equation (2b.13) holds whatever choice for B_{ref} is made, as long as $\mathrm{tr}_{\mathfrak{B}} B_{\mathrm{ref}}=1$. The formalism thus doesn't tell us how to choose this parameter. What we do see, however, from Eq. (2b.18), is that the influence of $\mathfrak{B}$ on $\mathfrak{S}$ expresses itself in terms of correlation functions $\mathrm{tr}_{\mathfrak{B}} B_1(t_1) B_2(t_2) \ldots B_n(t_n) B_{\mathrm{ref}}$. Here the B_i are the observables of $\mathfrak{B}$ occurring in the interaction Hamiltonian. Their time evolution is due to the free motion of $\mathfrak{B}$ governed by the evolution operator $\exp[-iL_{\mathfrak{B}}t]$. The state in which these correlation functions are to be evaluated is the parameter B_{ref}. This latter thus plays the role of a reference state for $\mathfrak{B}$. Therefore, the selection of the reference state should be based on the physics of a given system $\mathfrak{S} \oplus \mathfrak{B}$ and not much can be said in general. More specific statements are possible, e.g., for two special classes of problems. (i) If $\mathfrak{B}$ is a large system in equilibrium interacting so weakly with a small system $\mathfrak{S}$ that the equilibrium of $\mathfrak{B}$ is hardly disturbed then a reasonable choice for B_{ref} is, of course, the unperturbed equilibrium density operator of $\mathfrak{B}$. (ii) If $\mathfrak{S}$ and $\mathfrak{B}$ influence each other strongly – the expansions (2b.18) being possible though – and if the behavior of $\mathfrak{S}$ near a stationary regime is to be investigated, then the best possible reference state B_{ref} for $\mathfrak{B}$ can be evaluated selfconsistently in the following way. First $\varrho(t)$ is calculated with the expansions (2b.18) truncated at some order n with B_{ref} unspecified. Then, using (2b.12), $(1-\mathfrak{P})\,W(t)$ is evaluated in terms of $\varrho(t)$ with the expansions for the exponentials in (2b.12) carried up to order n. The stationary density operator of $\mathfrak{B}$ is then obtained as $\bar{B} = \mathrm{tr}_{\mathfrak{S}}\{\mathfrak{P}\,W(t\to\infty) + (1-\mathfrak{P})\,W(t\to\infty)\}$. Identifying $\bar{B}$ with B_{ref} one obtains an equation for the reference state of $\mathfrak{B}$. If such a procedure is carried out, the density operator $\varrho(t)$ of $\mathfrak{S}$ as well as the stationary density operator $\bar{B}$ of $\mathfrak{B}$ are determined by a selfconsistent systematic perturbation scheme. – Let us stress again that a physically well-motivated choice of the reference state B_{ref} of $\mathfrak{B}$ is not required for the generalized master equation (2b.13) to hold. It is, however, necessary to make a good choice for B_{ref} in order for the theory constituted by (2b.12), (2b.18) to give a reasonable description of the physical processes in question in low-order approximations of the expansions (2b.18) – if a low-order description is possible at all.

2c) Subsystems $\mathfrak{S}$ of Open Systems $\mathfrak{S} \oplus \mathfrak{B}$

$\mathfrak{S} \oplus \mathfrak{B}$ may be an open system since there are external time-dependent fields acting on it. Then the Liouville-von Neumann equation (2b.1) is replaced with

$$\dot{W}(t) = -(i/\hbar)\,[H(t), W(t)] = -iL(t)\,W(t)\,. \tag{2c.1}$$

That is, the Liouvillian $L(t)$ displays an explicit time dependence. Since the commutator $[L(t), L(t')]$ need not vanish we have to introduce a time-ordered product in Eq. (2b.12) which thus becomes

$$\begin{aligned}(1-\mathfrak{P})\, W(t) = & -i\, T\exp\left\{-i\int_0^t \mathrm{d}t'(1-\mathfrak{P})\, L(t')\right\}(1-\mathfrak{P})\, W(0) \\ & -\int_0^t \mathrm{d}t'\, T\exp\left\{-i\int_{t'}^t \mathrm{d}t''(1-\mathfrak{P})\, L(t'')\right\}(1-\mathfrak{P})\, L(t')\, \mathfrak{P} W(t')\,.\end{aligned} \tag{2c.2}$$

Here T is the usual time-ordering operator. The Nakajima-Zwanzig equation now reads

$$\begin{aligned}\dot{\varrho}(t) &= -i L_{\mathrm{eff}}(t)\, \varrho(t) + \int_0^t \mathrm{d}t'\, K(t,t')\, \varrho(t') + I(t) \\ L_{\mathrm{eff}}(t) &= \mathrm{tr}_{\mathfrak{B}}\, L(t)\, B_{\mathrm{ref}} \\ K(t,t') &= -\mathrm{tr}_{\mathfrak{B}}\, L(t)\, T\exp\left\{-i\int_{t'}^t \mathrm{d}t''(1-\mathfrak{P})\, L(t'')\right\}(1-\mathfrak{P})\, L(t')\, B_{\mathrm{ref}} \\ I(t) &= -i\,\mathrm{tr}_{\mathfrak{B}}\, L(t)\, T\exp\left\{-i\int_0^t \mathrm{d}t'(1-\mathfrak{P})\, L(t')\right\}(1-\mathfrak{P})\, W(0)\,.\end{aligned} \tag{2c.3}$$

We will need this version of the generalized master equation in Section 2e in order to calculate correlation functions of observables of $\mathfrak{S}$.

On the other hand, $\mathfrak{S}\oplus\mathfrak{B}$ may be an open system moving irreversibly under the influence of some other system $\mathfrak{C}$. The motion of $\mathfrak{S}\oplus\mathfrak{B}$ may then, of course, be described by a Nakajima-Zwanzig equation of the form (2b.13). We here consider – in view of the applications to be given later – the simple case where $\mathfrak{C}$ is a large system characterized by relaxation times much shorter than those of $\mathfrak{S}\oplus\mathfrak{B}$. In such a case we may neglect retardation effects in the Nakajima-Zwanzig equation for the density operator of $\mathfrak{S}\oplus\mathfrak{B}$ – which we keep denoting by $W(t)$ – and thus obtain the Markovian master equation

$$\dot{W}(t) = -i\Lambda\, W(t) \tag{2c.4}$$

with

$$\Lambda = L_{\mathrm{eff}} + i\int_0^\infty \mathrm{d}t\, K(t)\,. \tag{2c.5}$$

The quantities L_{eff} and $K(t)$ are given by (2b.14) with the symbols occurring there appropriately reinterpreted: L is the Liouvillian of $\mathfrak{S}\oplus+\mathfrak{B}\oplus\mathfrak{C}$, $\mathfrak{P}=C_{\mathrm{ref}}\,\mathrm{tr}_{\mathfrak{C}}$ is the projector used to eliminate the coordinates of $\mathfrak{C}$ with C_{ref} as the reference state for $\mathfrak{C}$. It is easily checked that the Liouvillian Λ generating the time evolution of the density operator $W(t)$ of $\mathfrak{S}\oplus\mathfrak{B}$, although no longer being of the form $\Lambda=[H,\ldots]$, conserves probability, i.e.

$$\mathrm{tr}_{\mathfrak{S}\oplus\mathfrak{B}}\Lambda = 0\,. \tag{2c.6}$$

We now assume that only the subsystem $\mathfrak{S}$ of $\mathfrak{S}\oplus\mathfrak{B}$ is of experimental relevance and thus construct the equation of motion for $\varrho(t)=\mathrm{tr}_{\mathfrak{B}}\,W(t)$. By going through the arguments of Section 2b again we find that $\varrho(t)$ obeys the Nakajima-Zwanzig equation (2b.13) with $\Lambda=\Lambda_{\mathfrak{S}}+\Lambda_{\mathfrak{B}}+\Lambda_{\mathfrak{S}\mathfrak{B}}$ as Liouvillian instead of L. This is so since at no point in Section 2b we have made use of any properties of L other than $\mathrm{tr}\,L=0$, $\mathrm{tr}_{\mathfrak{B}}\,L_{\mathfrak{B}}=0$ and $\mathrm{tr}_{\mathfrak{B}}\,L_{\mathfrak{S}}=L_{\mathfrak{S}}\,\mathrm{tr}_{\mathfrak{B}}$ which Λ has, too.

2d) Quasiprobability Distribution Functions

For some applications of the Nakajima-Zwanzig equation it is advantageous to describe the state of the system by a quasiprobability distribution of a complete set of observables rather than by a density operator. The observables are then represented by c-numbers variables[1] $S_1, S_2, \ldots B_1, B_2, \ldots$ corresponding to operators $\hat{S}_1, \hat{S}_2, \ldots, \hat{B}_1, \hat{B}_2, \ldots$ such that quantum-mechanical expectation values are given as moments of the quasiprobability distribution function $W(S_i, B_i, t)$

$$\begin{aligned}\langle\ldots\hat{S}_i\ldots\hat{B}_j\ldots\rangle &= \mathrm{tr}_{\mathfrak{S}\oplus\mathfrak{B}}\ldots\hat{S}_i\ldots\hat{B}_j\ldots\hat{W}(t)\\ &= \mathfrak{J}_{\mathfrak{S}\oplus\mathfrak{B}}\ldots S_i\ldots B_j\ldots W(S_i, B_j, t)\,.\end{aligned} \tag{2d.1}$$

Here the symbol

$$\mathfrak{J}_{\mathfrak{S}\oplus\mathfrak{B}}=\mathfrak{J}_{\mathfrak{S}}\mathfrak{J}_{\mathfrak{B}} \tag{2d.2}$$

denotes an integration over all variables S_i and B_i. It is always possible and can be very convenient to use such a c-number description since it sometimes reveals a physical process in question to be closely related to some classical random process. The c-number formalism can thereby help to gain physical insight and to find an adequate approximate treatment for the system considered. For detailed discussions of the mathematical properties of quasiprobability distributions we refer the reader to [9–14]. If the density operator $\hat{W}$ obeys a first order equation of motion in time, the associated quasiprobability distribution $W(S_i, B_i, t)$ can be defined such as to do so, too [15]. We then have

$$\dot{W}(S_i, B_i, t) = -i L W(S_i, B_i, t)\,, \tag{2d.3}$$

where L is a differential operator with respect to the variables S_i, B_i.

[1] Whenever necessary, we distinguish operators and associated c-number variables by the circonflex on the operator symbol.

This "Liouvillian" will have the properties

$$L = L_{\mathfrak{S}}(S_i) + L_{\mathfrak{B}}(B_i) + L_{\mathfrak{S}\mathfrak{B}}(S_i, B_i)$$
$$\mathfrak{I}_{\mathfrak{S}\oplus\mathfrak{B}} L = \mathfrak{I}_{\mathfrak{S}} L_{\mathfrak{S}} = \mathfrak{I}_{\mathfrak{B}} L_{\mathfrak{B}} = 0 \tag{2d.4}$$
$$\mathfrak{I}_{\mathfrak{B}} L_{\mathfrak{S}} = L_{\mathfrak{S}} \mathfrak{I}_{\mathfrak{B}} .$$

We now assume that only the variables S_i are of direct experimental relevance and therefore consider the reduced quasiprobability distribution function

$$\varrho(S_i, t) = \mathfrak{I}_{\mathfrak{B}} W(S_i, B_i, t) . \tag{2d.5}$$

It is easy to see, using (2d.4), that Eq. (2d.3) implies a Nakajima-Zwanzig equation for $\varrho(S_i, t)$. This latter reads like Eq. (2b.13) but all symbols occurring there have to be appropriately reinterpreted. Especially, the projector now is $\mathfrak{P} = B_{\text{ref}}(B_i)\, \mathfrak{I}_{\mathfrak{B}}$ and the reference state $B_{\text{ref}}(B_i)$ is a quasiprobability distribution for the variables B_i.

The c-number formalism just sketched becomes especially useful under the following circumstances. Suppose the separation of the set of variables $\{S_i, B_i\}$ in two subsets $\{S_i\}$ and $\{B_i\}$ does not correspond to a separation of the physical system $\mathfrak{S} \oplus \mathfrak{B}$ in two subsystems $\mathfrak{S}$ and $\mathfrak{B}$ with Hilbert spaces $\mathfrak{H}_{\mathfrak{S}}$ and $\mathfrak{H}_{\mathfrak{B}}$, respectively. It is then impossible to define a reduced density operator $\hat{\varrho}(t) = \text{tr}_{\mathfrak{B}} \hat{W}(t)$, since the partial trace $\text{tr}_{\mathfrak{B}}$ cannot be defined. It is possible, however, to construct the reduced quasiprobability distribution function by (2d.5). We will make use of this fact in Section 7.

2e) Correlation Functions of Observables of $\mathfrak{S}$

We here want to show how correlation functions of the observables S_i of the open system $\mathfrak{S}$ can be evaluated once the Nakajima-Zwanzig equation (2b.13) for the reduced density operator $\varrho(t)$ is solved. To this end we follow Haken and Weidlich [16] and couple the system $\mathfrak{S}$ to fictitious time-dependent external fields $\alpha_i(t)$, $\gamma_i(t)$ such that the Liouvillian becomes

$$L(t) = L + l(\alpha, \gamma, t) . \tag{2e.1}$$

There $L = L_{\mathfrak{S}} + L_{\mathfrak{B}} + L_{\mathfrak{S}\mathfrak{B}}$ is the time-independent Liouvillian encountered before. The additional term $l(\alpha, \gamma, t)$ is defined to act on some operator X as

$$l(\alpha, \gamma, t) X = \sum_i \{\alpha_i(t)\, S_i X - \gamma_i(t)\, X S_i\} . \tag{2e.2}$$

Since the density operator of $\mathfrak{S}$ may now be written as[2]

$$\begin{aligned}\varrho(\alpha,\gamma,t) &= \mathrm{tr}_{\mathfrak{B}}\, T\exp\left\{-i\int_0^t \mathrm{d}t'\, L(t')\right\} W(0)\\ &= \mathrm{tr}_{\mathfrak{B}}\left[T\exp\left\{-i\int_0^t \mathrm{d}t'\, H(t') - i\sum_i \int_0^t \mathrm{d}t'\,\alpha_i(t')\, S_i\right\}\right]\\ &\quad\cdot W(0)\left[\tilde{T}\exp\left\{+i\int_0^t \mathrm{d}t'\, H(t') + i\sum_i\int_0^t \mathrm{d}t'\,\gamma_i(t')\, S_i\right\}\right],\end{aligned} \tag{2e.3}$$

we have the following identity for correlation functions of the S_i

$$\begin{aligned}K_{nn'}(t_i,t_i') &= \langle S_{i_1'}(t_1')\, S_{i_2'}(t_2')\ldots S_{i_{n'}'}(t_{n'}')\, S_{i_n}(t_n)\ldots S_{i_2}(t_2)\, S_{i_1}(t_1)\rangle\\ &= \frac{1}{(-i)^n (i)^{n'}}\,\frac{\delta^n}{\delta\alpha_{i_1}(t_1)\ldots\delta\alpha_{i_n}(t_n)}\,\frac{\delta^{n'}}{\delta\gamma_{i_1'}(t_1')\ldots\delta\gamma_{i_{n'}'}(t_{n'}')}\\ &\quad \mathrm{tr}_{\mathfrak{S}}\,\varrho(\alpha,\gamma,t)\Big|_{\alpha,\gamma=0}.\end{aligned} \tag{2e.3}$$

We have written down this expression for the case $t_i \leqq t_{i+1}$, $t_i' \leqq t_{i+1}'$. More general time orderings can be treated correspondingly. The symbols $\delta/\delta\alpha(t)$ and $\delta/\delta\gamma(t)$ denote variational derivatives. As a next step we exploit the fact that the fictitious fields $\alpha_i(t)$ and $\gamma_i(t)$ enter the identity (2e.3) for the $n+n'$ points of time t_i, t_i' only. Without loss of generality we may therefore simplify the time-dependent part $l(\alpha,\gamma,t)$ of the Liouvillian $L(t)$ to

$$\begin{aligned} l(\alpha,\gamma,t) &= \sum_{\mu=1}^{n+n'} \delta(t-t_\mu)\, l_\mu\\ l_\mu X &= \begin{cases}\alpha_\mu S_\mu X & \text{for} \quad t_\mu = t_i\\ \gamma_\mu X S_\mu & \text{for} \quad t_\mu = t_i'.\end{cases}\end{aligned} \tag{2e.4}$$

Then the variational derivatives in Eq. (2e.3) are replaced with partial derivatives with respect to the parameters α_μ, γ_μ,

$$K_{nn'}(t_i,t_i') = \frac{1}{(-i)^n(i)^{n'}}\,\frac{\partial^{n+n'}}{\partial\alpha_{i_1}\partial\alpha_{i_2}\ldots\partial\alpha_{i_n}\partial\gamma_{i_1'}\ldots\partial\gamma_{i_{n'}'}}\,\varrho(\alpha,\gamma,t)\Big|_{\alpha,\gamma=0}. \tag{2e.5}$$

In order to evaluate the correlation functions $K_{nn'}(t_i,t_i')$ we thus have to first calculate the generating functional $\varrho(\alpha,\gamma,t)$ which is the density operator of $\mathfrak{S}$ in the presence of the fictitious fields $\alpha_i(t)$ and $\gamma_i(t)$. We can determine $\varrho(\alpha,\gamma,t)$ by solving the Zwanzig-Nakajima equation (2c.3) with the time-dependent Liouvillian (2e.1). By accounting for the special

[2] The time ordering operators $T(\tilde{T})$ arrange operators from left to right (right to left) according to increasing time arguments.

structure (2c.4) of $l(\alpha,\gamma,t)$ the equation of motion for the generating functional may be written as, for $t_\mu < t < t_{\mu+1}$,

$$\dot{\varrho}(\alpha,\gamma,t) = -iL_{\text{eff}}\,\varrho(\alpha,\gamma,t) + \int_{t_\mu}^{t} dt'\,K(t-t')\,\varrho(\alpha,\gamma,t') + I_\mu(t)\,. \tag{2e.6}$$

In L_{eff} and $K(t)$ the fictitious fields α, γ don't occur. These quantities are thus given by Eqs. (2b.14). The inhomogeneities $I_\mu(t)$, however, do depend on the fictitious fields α, γ and read

$$\begin{aligned} I_\mu(\alpha,\gamma,t) = &-i\,\mathrm{tr}_{\mathfrak{B}}\, L_{\mathfrak{S}\mathfrak{B}}\, e^{-i(1-\mathfrak{P})L(t-t_\mu)}\, e^{-il_\mu}\, e^{-i(1-\mathfrak{P})L(t_\mu-t_{\mu-1})} \\ &\cdot e^{-il_{\mu-1}} \ldots e^{-il_1}\, e^{-i(1-\mathfrak{P})Lt_1}(1-\mathfrak{P})\,W(0) \\ &-i\int_0^{t_1} dt'\,\mathrm{tr}_{\mathfrak{B}}\, L_{\mathfrak{S}\mathfrak{B}}\, e^{-i(1-\mathfrak{P})L(t-t_\mu)}\, e^{-il_\mu}\, e^{-i(1-\mathfrak{P})L(t_\mu-t_{\mu-1})} \ldots \\ &\cdot e^{-il_1}\, e^{-i(1-\mathfrak{P})L(t_1-t')}(1-\mathfrak{P})\,LB_{\text{ref}}\,\varrho(\alpha,\gamma,t') \\ &-i\int_{t_1}^{t_2} dt'\,\mathrm{tr}_{\mathfrak{B}}\, L_{\mathfrak{S}\mathfrak{B}}\, e^{-i(1-\mathfrak{P})L(t-t_\mu)}\, e^{-il_\mu}\, e^{-i(1-\mathfrak{P})L(t_\mu-t_{\mu-1})} \ldots \\ &\cdot e^{-il_2}\, e^{-i(1-\mathfrak{P})L(t_2-t')}(1-\mathfrak{P})\,LB_{\text{ref}}\,\varrho(\alpha,\gamma,t') \\ &\ldots\ldots \\ &-i\int_{t_{\mu-1}}^{t_\mu} dt'\,\mathrm{tr}_{\mathfrak{B}}\, L_{\mathfrak{S}\mathfrak{B}}\, e^{-i(1-\mathfrak{P})L(t-t_\mu)}\, e^{-il_\mu} \\ &\cdot e^{-i(1-\mathfrak{P})L(t_\mu-t')}(1-\mathfrak{P})\,LB_{\text{ref}}\,\varrho(\alpha,\gamma,t') \end{aligned} \tag{2e.7}$$

and, for $\mu = 0$,

$$I_0(t) \equiv I(t) = -i\,\mathrm{tr}_{\mathfrak{B}}\, L_{\mathfrak{S}\mathfrak{B}}\, e^{-i(1-\mathfrak{P})Lt}(1-\mathfrak{P})\,W(0)\,.$$

In bringing the Nakajima-Zwanzig equation (2c.3) to the form (2c.6) we have used the time ordering prescription to arrange all operators properly with $t_\nu \leqq t_{\nu+1}$. We also have carried out some time integrals according to

$$\int_{t_\nu}^{t_{\nu+1}} dt\,L = (t_{\nu+1} - t_\nu)\,L$$

and

$$\int_{t_\nu-0}^{t_\nu+0} l(\alpha,\gamma,t)\,dt = l_\nu\,.$$

Because of $\mathfrak{P}(1-\mathfrak{P}) = 0$ and, by the definition of l_ν, $\mathfrak{P}\,l_\nu = l_\nu\,\mathfrak{P}$, we have replaced $(1-\mathfrak{P})\,l_\nu$ by l_ν everywhere. It is interesting to note that the inhomogeneity $I_\mu(t)$ contains the generating functional $\varrho(\alpha,\gamma,t)$ for preceding time intervals, i.e. for $t < t_\mu$. These terms may be looked upon as being determined by solving Eq. (2e.6) for the preceding time intervals.

As indicated explicitly, Eq. (2e.6) governs the behavior of the generating functional between the "jumps" at the times t_μ and $t_{\mu+1}$. At these times $\varrho(\alpha, \gamma, t)$ changes discontinuously according to the δ-functions in $l(\alpha, \gamma, t)$. It is important to realize that we need to know the magnitudes of the jumps of $\varrho(\alpha, \gamma, t)$ to first order in the parameters α_ν, γ_ν only. Higher order terms do not contribute to the correlation functions $K_{nn'}(t_i, t_i')$ as is seen from Eq. (2e.5). To first order, the jump is determined by the first term in the right member of Eq. (2c.3), namely by

$$L_{\text{eff}}(t) = L_{\text{eff}} + l(\alpha, \gamma, t)\,.$$

By using Eq. (2e.4) we find

$$\varrho(\alpha, \gamma, t_\nu + 0) = (1 - i\, l_\nu)\, \varrho(\alpha, \gamma, t_\nu - 0)\,. \tag{2e.8}$$

The problem formulated by Eqs. (2e.6) and (2e.8) is to integrate (2e.6) stepwise from jump to jump accounting for a new initial condition (2e.8) at each t_μ. This program can be carried out formally using the time evolution operator $V(t)$ defined in (2b.17). For $t > t_\mu$ we obtain

$$\begin{aligned} \varrho(\alpha, \gamma, t) &= V(t - t_\mu)\, \varrho(\alpha, \gamma, t_\mu + 0) + \int_{t_\mu}^{t} \mathrm{d}t'\, V(t - t')\, I_\mu(t') \\ &= V(t - t_\mu)\,(1 - i\, l_\mu)\, \varrho(\alpha, \gamma, t_\mu - 0) + \int_{t_\mu}^{t} \mathrm{d}t'\, V(t - t')\, I_\mu(t')\,. \end{aligned} \tag{2e.9}$$

By putting together the solutions for all preceding time intervals we finally get

$$\begin{aligned} \varrho(\alpha, \gamma, t) = {} & V(t - t_\mu)\,(1 - i\, l_\mu)\, V(t_\mu - t_{\mu-1})\,(1 - i\, l_{\mu-1}) \ldots V(t_2 - t_1) \\ & \cdot (1 - i\, l_1)\, V(t_1)\, \varrho(0) \\ & + V(t - t_\mu)\,(1 - i\, l_\mu)\, V(t_\mu - t_{\mu-1})\,(1 - i\, l_{\mu-1}) \ldots V(t_2 - t_1) \\ & \cdot (1 - i\, l_1) \int_0^{t_1} \mathrm{d}t'\, V(t_1 - t')\, I_0(t') \\ & + V(t - t_\mu)\,(1 - i\, l_\mu)\, V(t_\mu - t_{\mu-1})\,(1 - i\, l_{\mu-1}) \ldots V(t_3 - t_2) \\ & \cdot (1 - i\, l_2) \int_{t_1}^{t_2} \mathrm{d}t'\, V(t_2 - t')\, I_1(t') \\ & + \cdots \\ & + V(t - t_\mu)\,(1 - i\, l_\mu) \int_{t_{\mu-1}}^{t_\mu} \mathrm{d}t'\, V(t_\mu - t')\, I_{\mu-1}(t') \\ & + \int_{t_\mu}^{t} \mathrm{d}t'\, V(t - t')\, I_\mu(t')\,. \end{aligned} \tag{2e.10}$$

This expression is not yet fully explicit since, as mentioned above, the inhomogeneities $I_\nu(t)$ contain $\varrho(\alpha, \gamma, t)$ of preceding time intervals. We refrain, however, from writing down the fully explicit expression as its length makes it even less enlightening than the one given above. In discussing (2e.10) we remark that the first term occuring on the right hand side is distinguished from all the others. It would be the only one to appear in the trivial special case of no interaction between $\mathfrak{S}$ and $\mathfrak{B}$ since, according to Eq. (2e.7), all $I_\mu(t)$ vanish identically for $L_{\mathfrak{S}\mathfrak{B}}=0$. Then the time evolution operator $V(t)$ of course degenerates to the unitary operator $\exp(-i L_{\mathfrak{S}} t)$. There is another special case in which all the $I_\mu(t)$ vanish or rather are negligible, namely if $\mathfrak{S}$ undergoes a Markovian motion under the influence of $\mathfrak{B}$ [16]. We shall show this below.

It is now a straightforward matter to carry out the prescription (2e.5) on $\varrho(\alpha, \gamma, t)$ as given by (2e.10) and to write down the expression for the correlation function $K_{nn'}(t_i, t_i')$. We will not do that for the general $K_{nn'}(t_i, t_i')$ since the resulting formula is lengthy and will hardly ever be needed. Let us rather illustrate the steps to be taken for the important special case of equilibrium correlation functions. In equilibrium we have

$$
\begin{gathered}
L W(0)=0 \\
\text{and} \\
\varrho(t)=V(t)\, \varrho(0)+\int_0^t \mathrm{d}t'\, V(t-t')\, I_0(t')=\varrho(0)\,.
\end{gathered}
\tag{2e.11}
$$

For the simplest two-times correlation function we then obtain, for $t_2-t_1 \geqq 0$,

$$
\begin{aligned}
\langle S_2(t_2)\, S_1(t_1)\rangle &= \frac{1}{(-i)^2}\, \frac{\partial^2}{\partial\alpha_1 \partial\alpha_2}\, \mathrm{tr}_{\mathfrak{S}}\, \varrho(\alpha, \gamma, t)\bigg|_{\alpha,\gamma=0} \\
&= \frac{\partial^2}{\partial\alpha_1 \partial\alpha_2}\, \mathrm{tr}_{\mathfrak{S}} \left\{ l_2\, V(t_2-t_1)\, l_1\, \varrho(0) + l_2 \int_{t_1}^{t_2} \mathrm{d}t'\, V(t_2-t')\, I_1(t') \right\}\bigg|_{\alpha,\gamma=0} \\
&= \mathrm{tr}_{\mathfrak{S}}\, S_2\, V(t_2-t_1)\, S_1\, \varrho(0) \\
&\quad -i \int_0^{t_2-t_1} \mathrm{d}t'\, \mathrm{tr}_{\mathfrak{S}}\, S_2\, V(t_2-t_1-t')\, \mathrm{tr}_{\mathfrak{B}}\, L_{\mathfrak{S}\mathfrak{B}}\, \mathrm{e}^{-i(1-\mathfrak{P})Lt'}\, S_1 (1-\mathfrak{P})\, W(0)\,.
\end{aligned}
\tag{2e.12}
$$

Correspondingly, again for $t_2-t_1 \geqq 0$,

$$
\begin{aligned}
\langle S_1(t_1)\, S_2(t_2)\rangle &= \frac{\partial^2}{\partial\alpha_2 \partial\gamma_1}\, \mathrm{tr}_{\mathfrak{S}}\, \varrho(\alpha, \gamma, t)\bigg|_{\alpha,\gamma=0} \\
&= \mathrm{tr}_{\mathfrak{S}}\, S_2\, V(t_2-t_1)\, \varrho(0)\, S_1 \\
&\quad -i \int_0^{t_2-t_1} \mathrm{d}t'\, \mathrm{tr}_{\mathfrak{S}}\, S_2\, V(t_2-t_1-t')\, \mathrm{tr}_{\mathfrak{B}}\, L_{\mathfrak{S}\mathfrak{B}}\, \mathrm{e}^{-i(1-\mathfrak{P})Lt'} (1-\mathfrak{P})\, W(0)\, S_1\,.
\end{aligned}
\tag{2e.13}
$$

As a last example we consider the four-time correlation function

$$\langle S_{1'}(t'_{1'})\, S_{2'}(t'_{2'})\, S_2(t_2)\, S_1(t_1)\rangle = \frac{\partial^4}{\partial\alpha_1\partial\alpha_2\partial\gamma_{1'}\partial\gamma_{2'}}\operatorname{tr}_{\mathfrak{S}}\varrho(\alpha,\gamma,t)\Big|_{\alpha,\gamma=0} \tag{2e.14}$$

where $t'_2 \geqq t'_1$, $t_2 \geqq t_1$. Since such correlation functions with "pyramidal" time order are encountered in quantum optics mainly we evaluate (2e.14) for the special case of the so called intensity correlation function for a mode of an electromagnetic field.

$$\begin{aligned}\langle b^\dagger(t)\, b^\dagger(t+\tau)\, b(t+\tau)\, b(t)\rangle &= \operatorname{tr}_{\mathfrak{S}} b[V(\tau)\, b\varrho(0)\, b^\dagger]\, b^\dagger \\ &- i\int_0^\tau \mathrm{d}t' \operatorname{tr}_{\mathfrak{S}} b[V(\tau-t')\operatorname{tr}_{\mathfrak{B}} L_{\mathfrak{S}\mathfrak{B}}\mathrm{e}^{-i(1-\mathfrak{P})Lt'} b(1-\mathfrak{P})\, W(0)\, b^\dagger]\, b^\dagger\,.\end{aligned} \tag{2e.15}$$

Here b and $b^\dagger$ are the annihilation and creation operator of photons in the field mode $\mathfrak{S}$ considered. The surroundings $\mathfrak{B}$ of the field mode is, for a laser, constituted by active atoms and pump and loss mechanisms.

In case of need other correlation functions can be constructed analogously. The resulting expressions become rather lengthy for $n+n'$ increasing. However, in the above-mentioned special case of a Markovian motion of $\mathfrak{S}$ even the general formula for $K_{nn'}(t_i, t'_i)$ is easily written down and has a rather compact appearance. To see the simplifications then possible we first discuss the behavior of the time evolution operator $V(t)$ which obeys the equation of motion (2b.17),

$$\dot{V}(t) = -iL_{\text{eff}}\, V(t) + \int_0^t \mathrm{d}t'\, K(t-t')\, V(t')\,.$$

As already discussed in subsection 2b – following Eq. (2b.18) – the time dependence of the integral kernel is determined by that of unperturbed correlation functions of observables of $\mathfrak{B}$. Now if $\mathfrak{B}$ is a large system with internal relaxation times $\tau_{\mathfrak{B}}$ very short compared to the relaxation times $\tau_{\mathfrak{S}}$ of $\mathfrak{S}$ (i.e. of $V(t)$), we can, for times $t \gg \tau_{\mathfrak{B}}$ neglect retardation effects in the equation of motion for $V(t)$ and thus have

$$\dot{V}(t) = \Lambda V(t) \quad \text{or} \quad V(t) = \mathrm{e}^{\Lambda t} \tag{2e.16}$$

with

$$\Lambda = -iL_{\text{eff}} + \int_0^\infty \mathrm{d}t\, K(t)\,.$$

By a similar argument one shows that the inhomogeneities $I_\mu(t)$ in Eq. (2e.7) are manifestations of memory effects and vanish in the Markovian limit $\tau_{\mathfrak{S}}/\tau_{\mathfrak{B}} \approx 0$. Then we immediately find from Eqs. (2e.5) and (2e.10)

$$\begin{aligned}
&\langle S_1'(t_1)\, S_2'(t_2) \ldots S_n'(t_n)\, S_n(t_n)\, S_{n-1}(t_{n-1}) \ldots S_2(t_2)\, S_1(t_1)\rangle \\
&= \mathrm{tr}_{\mathfrak{S}}\, S_n\, V(t_n - t_{n-1})\, [S_{n-1}\, V(t_{n-1} - t_{n-2}) \qquad (2e.17)\\
&\quad \cdot [S_{n-2} \ldots S_2\, V(t_2 - t_1)\, [S_1\, V(t_1)\, \varrho(0)\, S_1']\, S_2'] \ldots S_{n-2}']\, S_{n-1}']\, S_n' .
\end{aligned}$$

This quite wellknown expression [16–21] may be looked upon as a generalized fluctuation-dissipation theorem, valid for quantum mechanical Markov processes. It relates the mean irreversible motion of $\mathfrak{S}$ – as characterized by the time evolution operator $V(t)$ – to fluctuations in $\mathfrak{S}$ – as expressed by correlation functions of observables of $\mathfrak{S}$.

Let us conclude with a few remarks on where and how the above expressions for the correlation functions $K_{nn'}(t_i, t_i')$ can be used. One application is obvious. If for a given system the Nakajima-Zwanzig equation is solved, that is if the time evolution operator $V(t)$ is known, the above results allow an explicit evaluation of the $K_{nn'}(t_i, t_i')$. On the other hand, the formal expressions for the $K_{nn'}$ can be used to construct hierarchies of equations of motion for correlation functions or Green's functions. Such hierarchies may be easier to solve than the Nakajima-Zwanzig equation itself. We shall illustrate this use of the above results in section 4 in our treatment of superconductivity.

3. Linear Damping Phenomena

3a) Introductory Remarks

We here want to illustrate the applicability of the Nakajima-Zwanzig theory to damping phenomena in microscopic systems as produced by a weak coupling to large systems in thermal equilibrium. As already stated in Section 1, the first such application was made by Argyres and Kelley [8] in a treatment of spin relaxation. We will briefly review their results in subsection 3c. The main body of this section, subsection 3b, will be concerned with an even simpler but no less important case, the damped harmonic oscillator.

Linear damping phenomena can be and have been treated by other methods as well. Wangsness and Bloch [22] have treated spin relaxation by constructing and solving a master equation for the density operator of the spin system. Their investigation follows the lines suggested by Pauli [1] and especially uses Pauli's assumption of repeatedly random phases which we have discussed in Section 1. In the same spirit the damped harmonic oscillator has been dealt with by Weidlich and Haake [23].

The advantage of the modern theory of damping phenomena using the techniques of Nakajima and Zwanzig over the older theories using Pauli's method is twofold. First, the theoretically unsatisfying assumption of repeatedly random phases can be avoided and second, more complicated phenomena like non-Markovian damping effects, inaccessible to Pauli's method, can be handled quite easily. – There is another way of dealing with linear damping phenomena, first laid out by Senitzky [24] and later generalized by Mori [25]. These authors describe the dynamics of the open system $\mathfrak{S}$ and the heat bath $\mathfrak{B}$ in the Heisenberg picture. By eliminating the observables of $\mathfrak{B}$ from the Heisenberg equations of motion for the observables of $\mathfrak{S}\oplus\mathfrak{B}$ quantum mechanical Langevin equations for the observables of $\mathfrak{S}$ are obtained. These methods will not be considered in the present paper. We want to emphasize, however, that they are equivalent to the Nakajima-Zwanzig method, just as Schrödinger picture and Heisenberg picture are equivalent.

3b) The Damped Harmonic Oscillator

We consider an ideal oscillator $\mathfrak{S}$ coupled to a heat bath $\mathfrak{B}$. The heat bath is required to have the following properties. (i) It is in thermal equilibrium before the interaction with the oscillator is switched on. (ii) It is a very large system with internal relaxation times $\tau_{\mathfrak{B}}$ very short compared to the relaxation time $\tau_{\mathfrak{S}}$ of the oscillator which is to be determined. (iii) It is sufficiently large and so weakly coupled to the oscillator that its thermal equilibrium is never disturbed appreciably by the oscillator. We shall first naively use and later discuss in some more detail these three conditions.

The Hamiltonian of the free oscillator is

$$H_{\mathfrak{S}} = \hbar\omega b^{\dagger} b \tag{3b.1}$$

where the (Bose) operators b and $b^{\dagger}$ annihilate and create, respectively, quanta of energy $\hbar\omega$ in the oscillator. The free heat bath Hamiltonian $H_{\mathfrak{B}}$ need not be specified. The interaction Hamiltonian $H_{\mathfrak{S}\mathfrak{B}}$ we choose as

$$H_{\mathfrak{S}\mathfrak{B}} = \hbar g(bB^{\dagger} + b^{\dagger} B) \tag{3b.2}$$

with unspecified dimensionless heat bath operators B and $B^{\dagger}$. The coupling constant g has the dimension of a frequency. More general couplings between $\mathfrak{S}$ and $\mathfrak{B}$ can and for some applications even have to be considered [18].

The density operator $\varrho(t)$ of the oscillator obeys the Nakajima-Zwanzig equation (2b.13). The reference state B_{ref} for the heat bath occuring there can be taken, because of condition (iii) above, as the unperturbed canonical density operator

$$B_{\text{ref}} = e^{-\beta H_{\mathfrak{B}}}/\mathrm{tr}_{\mathfrak{B}}\, e^{-\beta H_{\mathfrak{B}}} . \tag{3b.3}$$

Because of condition (ii) both the integral kernel $K(t)$ and the inhomogeneity $I(t)$ in Eq. (2b.13) relax on a time scale $\tau_{\mathfrak{B}}$ which is much shorter than the time scale $\tau_{\mathfrak{S}}$ characteristic for the motion of $\varrho(t)$. We therefore expect, for times $t \gg \tau_{\mathfrak{B}}$, a Markov approximation to the Nakajima-Zwanzig equation to hold

$$\begin{aligned} &\dot{\varrho}(t) = \Lambda \varrho(t) \\ &\quad \Lambda = -iL_{\mathfrak{S}} - i\,\mathrm{tr}_{\mathfrak{B}} L_{\mathfrak{S}\mathfrak{B}} B_{\text{ref}} + \int_0^\infty \mathrm{d}t\, K(t) . \end{aligned} \tag{3b.4}$$

It is easy to see that the term $\mathrm{tr}_{\mathfrak{B}} L_{\mathfrak{S}\mathfrak{B}} B_{\text{ref}}$ in Λ corresponds to a mean conservative force exerted on the oscillator by the heat bath. Such an effect is easily treated but of no interest in the present context. We therefore assume this mean force to vanish, i.e.

$$\mathrm{tr}_{\mathfrak{B}} B B_{\text{ref}} = 0 . \tag{3b.5}$$

Finally, we take condition (iii) above to imply that third and higher order contributions in g to the integral kernel $K(t)$ can be neglected and thus get

$$\Lambda = -iL_{\mathfrak{S}} - \int_0^\infty \mathrm{d}t\, \mathrm{tr}_{\mathfrak{B}} L_{\mathfrak{S}\mathfrak{B}} e^{-i(L_{\mathfrak{S}} + L_{\mathfrak{B}})t} L_{\mathfrak{S}\mathfrak{B}} B_{\text{ref}} . \tag{3b.6}$$

The evaluation of (3b.6) is a simple exercise. The resulting master equation for the oscillator reads

$$\begin{aligned} \dot{\varrho}(t) = &-i(\omega + \Delta)\, [b^\dagger b, \varrho(t)] \\ &+ \kappa\{[b, \varrho(t) b^\dagger] + [b\varrho(t), b^\dagger]\} \\ &+ 2\kappa \bar{n}[b, [\varrho(t), b^\dagger]] . \end{aligned} \tag{3b.7}$$

The influence of the heat bath on the oscillator is characterized by the three real parameters Δ, κ, and $\bar{n}$. These are found as

$$\begin{aligned} &\kappa + i\Delta = g^2 \int_0^\infty \mathrm{d}t\, e^{i\omega t} \langle [B(t), B^\dagger(0)] \rangle \\ &\quad \kappa \bar{n} = g^2 \,\mathrm{Re} \int_0^\infty \mathrm{d}t\, e^{i\omega t} \langle B^\dagger(0)\, B(t) \rangle , \end{aligned} \tag{3b.8}$$

where $\langle \cdots \rangle = \mathrm{tr}_{\mathfrak{B}} \ldots B_{\text{ref}}$ and $B(t) = e^{iH_{\mathfrak{B}}t/\hbar} B(0)\, e^{-iH_{\mathfrak{B}}t/\hbar}$. We see that the parameters $\Delta, \kappa, \bar{n}$ are given as Fourier transforms of retarded equilibrium Green's functions of the bath operators B and $B^\dagger$, the Fourier transforms being evaluated at the eigenfrequency ω of the ideal oscillator. As a formal remark we note that κ and Δ, as functions of the "variable" ω, are related by a dispersion relation. Moreover, and more importantly, the two Green's functions determining κ and $\kappa\bar{n}$ are related by the fluctuation dissipation theorem [26]. As a consequence,

$$\bar{n} = [e^{\beta\hbar\omega} - 1]^{-1}. \tag{3b.9}$$

The physical meaning of the parameters κ, Δ, and $\bar{n}$ becomes obvious when equations of motion for the oscillator amplitude and the mean number of quanta are extracted from Eq. (3b.7):

$$\begin{aligned} \langle \dot{b}(t)\rangle &= \{-i(\omega+\Delta)-\kappa\}\,\langle b(t)\rangle \\ \langle b^\dagger \dot{b}(t)\rangle &= -2\kappa\{\langle b^\dagger b(t)\rangle - \bar{n}\}\,. \end{aligned} \tag{3b.10}$$

We thus see that Δ is a frequency shift, κ a damping constant, and $\bar{n}$ the stationary number of quanta. According to Eq. (3b.9) the stationary number of quanta is what we would expect for thermal equilibrium at temperature $1/\beta$. By observing that the stationary solution of the master equation (3b.7) is[3]

$$\bar{\varrho} = e^{-\beta\hbar\omega b^\dagger b}/\mathrm{tr}_{\mathfrak{S}}\, e^{-\beta\hbar\omega b^\dagger b} \tag{3b.11}$$

we conclude that the heat bath imposes its thermal equilibrium at temperature $1/\beta$ on the oscillator.

The physical nature of the motion of the oscillator described by Eq. (3b.7) is most obviously displayed when this equation is rewritten in terms of Glauber's diagonal representation of $\varrho(t)$ with respect to coherent states [11]

$$\varrho(t) = \int d^2\beta\, P(\beta,\beta^*,t)\,|\beta\rangle\,\langle\beta| \quad \text{with} \quad b|\beta\rangle = \beta|\beta\rangle\,. \tag{3b.12}$$

This representation allows the computation of normally ordered expectation values $\langle b^{\dagger n} b^m\rangle$ as moments of the weight function $P(\beta,\beta^*,t)$ as

$$\langle b^\dagger(t)^n b(t)^m\rangle = \int d^2\beta\, \beta^{*n}\beta^m P(\beta,\beta^*,t)\,. \tag{3b.13}$$

By using the wellknown commutation relations

$$\begin{aligned} [b, f(b,b^\dagger)] &= \partial f(b,b^\dagger)/\partial b^\dagger \\ [f(b,b^\dagger), b^\dagger] &= \partial f(b,b^\dagger)/\partial b \end{aligned} \tag{3b.14}$$

[3] This is most easily verified in the representation in which the number operator $b^\dagger b$ is diagonal.

the master equation (3b.7) is easily transformed into the following differential equation of motion for the weight function $P(\beta, \beta^*, t)$ [18]

$$\begin{aligned}\dot{P}(\beta, \beta^*, t) = \bigg\{ & -i(\omega + \Delta)\left(\frac{\partial}{\partial \beta^*}\beta^* - \frac{\partial}{\partial \beta}\beta\right) \\ & + \kappa\left(\frac{\partial}{\partial \beta^*}\beta^* + \frac{\partial}{\partial \beta}\beta\right) \\ & + 2\kappa\bar{n}\frac{\partial^2}{\partial \beta^* \partial \beta}\bigg\} P(\beta, \beta^*, t) .\end{aligned} \tag{3b.15}$$

This is a Fokker Planck equation for a Gaussian Markov process [27]. It is known in the theory of classical random processes as the Fokker Planck equation of the damped harmonic oscillator [28]. Its solution is known as

$$P(\beta, \beta^*, t) = \int d^2\beta_0 P(\beta, \beta^*, t|\beta_0, \beta_0^*)\, P(\beta_0, \beta_0^*, 0) , \tag{3b.16}$$

where $P(\beta, \beta^*, 0)$ is the initial P-function and $P(\beta, \beta^*, t|\beta_0, \beta_0^*)$ a "transition probability"

$$P(\beta, \beta^*, t|\beta_0, \beta_0^*) = [\pi\bar{n}(1 - e^{-2\kappa t})]^{-1} \exp\left\{ - \frac{|\beta - \beta_0 e^{-(i\omega + i\Delta + \kappa)t}|^2}{\bar{n}(1 - e^{-2\kappa t})} \right\} . \tag{3b.17}$$

Let us now discuss the assumptions (ii) and (iii) stated at the beginning of this subsection. They can both be formulated in a somewhat more quantitative manner. First, the relevant time scale $\tau_{\mathfrak{B}}$ for the motion of $\mathfrak{B}$ is seen to be given by the decay times of the equilibrium Green's functions occuring in Eqs. (3b.8). The time scale for the oscillator is κ^{-1}, i.e. the inverse damping constant. The validity of the Markov approximation thus requires

$$\tau_{\mathfrak{B}}\kappa \ll 1 . \tag{3b.18}$$

Next, assumption (iii) has been used as a motivation to choose the thermal equilibrium density operator (3b.3) as a reference state for the heat bath $\mathfrak{B}$ and to approximate the integral kernel $K(t)$ to lowest order in the coupling constant g (Born approximation). As a result, we have seen the oscillator relax to the thermal equilibrium state (3b.11) at the temperature $1/\beta$ of the heat bath. In order to explicitly justify the Born approximation we would have to estimate the contribution of all higher order terms in $K(t)$ to Δ in Eq. (3b.4). Such an estimation cannot be carried out in general, that is unless a specific model for the heat bath is considered. What we can do without specifying detailed properties of the heat bath, however, is to perform a consistency check on our arguments.

For the above treatment of the damped harmonic oscillator to be meaningful the perturbation of the thermal equilibrium of the heat bath caused by the coupling to the oscillator must be negligible. The state of the heat bath is given by

$$\begin{aligned}\varrho_{\mathfrak{B}}(t) &= \mathrm{tr}_{\mathfrak{S}}\, W(t) = \mathrm{tr}_{\mathfrak{S}}\{\mathfrak{P}\, W(t) + (1-\mathfrak{P})\, W(t)\} \\ &= B_{\mathrm{ref}} + \mathrm{tr}_{\mathfrak{S}}\, \mathrm{e}^{-i(1-\mathfrak{P})Lt}(1-\mathfrak{P})\, W(0) \\ &\quad - i\,\mathrm{tr}_{\mathfrak{S}} \int_0^t \mathrm{d}t'\, \mathrm{e}^{-i(1-\mathfrak{P})Lt'}(1-\mathfrak{P})\, L B_{\mathrm{ref}}\, \varrho(t-t')\,.\end{aligned} \tag{3b.19}$$

In evaluating the deviation from thermal equilibrium, $\varrho_{\mathfrak{B}}(t) - B_{\mathrm{ref}}$, we have to make the same approximations used in determining the state $\varrho(t)$ of the oscillator. The second term on the right hand side of Eq. (3b.19) decays on a time scale $\tau_{\mathfrak{B}}$ and may thus be neglected for $t \gg \tau_{\mathfrak{B}}$. The third term reads, to order g^2,

$$\begin{aligned}\varrho_{\mathfrak{B}}(t) - B_{\mathrm{ref}} = &- i \int_0^t \mathrm{d}t'\, \mathrm{tr}_{\mathfrak{S}}\, \mathrm{e}^{-i(L_{\mathfrak{S}}+L_{\mathfrak{B}})t'} L_{\mathfrak{S}\mathfrak{B}} B_{\mathrm{ref}}\, \varrho(t-t') \\ &- \int_0^t \mathrm{d}t' \int_0^{t'} \mathrm{d}t''\, \mathrm{tr}_{\mathfrak{S}}\, \mathrm{e}^{-i(L_{\mathfrak{S}}+L_{\mathfrak{B}})t''} L_{\mathfrak{S}\mathfrak{B}}\, \mathrm{e}^{-i(L_{\mathfrak{S}}+L_{\mathfrak{B}})(t'-t'')} L_{\mathfrak{S}\mathfrak{B}} B_{\mathrm{ref}}\, \varrho(t-t')\,.\end{aligned} \tag{3b.20}$$

Let us consider the diagonal matrix elements

$$\Delta B_n(t) \equiv \langle n|\varrho_{\mathfrak{B}}|n\rangle - \langle n|B_{\mathrm{ref}}|n\rangle \quad \text{with} \quad H_{\mathfrak{B}}|n\rangle = E_n|n\rangle \tag{3b.21}$$

with respect to energy eigenstates. To these the first term in (3b.20) does not contribute because of (3b.5). The contribution of the second term is easily found, for $t \gg \tau_{\mathfrak{B}}$, to read[4]

$$\begin{aligned}&\Delta B_n(t) = -\langle n|B_{\mathrm{ref}}|n\rangle\, (\alpha_n/\kappa)\, (1-\mathrm{e}^{-2\kappa t})\, \frac{\langle b^\dagger b(0)\rangle - \bar{n}}{1+\bar{n}} \\ &\text{with} \\ &\langle n|B_{\mathrm{ref}}|n\rangle\, \alpha_n = g^2\, \mathrm{Re} \int_0^\infty \mathrm{d}t\, \mathrm{e}^{i\omega t} \langle n|[B(t), B^\dagger(0)]|n\rangle\,.\end{aligned} \tag{3b.22}$$

For the Born approximation to make sense the relative deviation from thermal equilibrium must be small, i.e.

$$\left|\frac{\Delta B_n(t)}{\langle n|B_{\mathrm{ref}}|n\rangle}\right| = \frac{|\alpha_n|}{\kappa}(1-\mathrm{e}^{-2\kappa t})\left|\frac{\langle b^\dagger(0)\, b(0)\rangle - \bar{n}}{1+\bar{n}}\right| \ll 1\,. \tag{3b.23}$$

In discussing this condition we will restrict ourselves to a few remarks. First, we see that the initial excitation $\langle b^\dagger(0)\, b(0)\rangle$ of the oscillator must

[4] To avoid irrelevant complications we here assume $\langle n|BB|n\rangle = 0$.

not be too high. Moreover, the coupling of $\mathfrak{S}$ and $\mathfrak{B}$ must be such that $|\alpha_n|/\kappa \ll 1$. To appreciate this let us write down α_n in the $H_{\mathfrak{B}}$-representation.

$$\begin{aligned} &\alpha_n = \alpha_n' - \alpha_n'' \\ &\alpha_n' = \pi \hbar g^2 \sum_m \delta(\hbar\omega + E_n - E_m)\, |\langle n|B|m\rangle|^2 \\ &\alpha_n'' = \mathrm{e}^{\beta\hbar\omega} \pi \hbar g^2 \sum_m \delta(\hbar\omega + E_m - E_n)\, |\langle m|B|n\rangle|^2 . \end{aligned} \tag{3b.24}$$

By comparison of (3b.22) with (3b.8) we also find

$$\sum_n \langle n|B_{\mathrm{ref}}|n\rangle\, \alpha_n' = \sum_n \langle n|B_{\mathrm{ref}}|n\rangle\, \alpha_n'' = \kappa(1 - \mathrm{e}^{-\beta\hbar\omega})^{-1} . \tag{3b.25}$$

These expressions suggest to interpret the α_n as transition rates between bath states of energies E_n and E_m. α_n' measures the rate of change of the occupation probability of bath states with energy E_n due to transitions $n \to m$ to states with energy $E_m = E_n + \hbar\omega$. Likewise, α_n'' accounts for transitions $m \to n$ from states with energy $E_m = E_n - \hbar\omega$. We may thus say that the smallness of α_n/κ implies that none of the contributions $\langle n|B_{\mathrm{ref}}|n\rangle\, \alpha_n$ to the damping constant κ exhaust their sum. For more detailed conclusions we refer to [29].

3c) Spin Relaxation

The discussion of spin relaxation is precisely analogous to that of the damped oscillator given above. Therefore and since the literature abounds of detailed presentations of spin relaxation theory [8, 22, 30] we here merely write down what we will need in Section 6, namely the master equation for a damped spin-$\frac{1}{2}$ system. It reads, if both the Markov and the Born approximations are made and with the energy shifts Δ oppressed,

$$\begin{aligned} \dot{\varrho}(t) = &-(i/\hbar)\,[H_{\mathfrak{S}}, \varrho(t)] \\ &+ \tfrac{1}{2}\gamma_{10}\{s^-, \varrho(t)\, s^+] + [s^- \varrho(t), s^+]\} \\ &+ \tfrac{1}{2}\gamma_{01}\{[s^+, \varrho(t)\, s^-] + [s^+ \varrho(t), s^-]\} \\ &+ \tfrac{1}{2}\eta\,\{[s^z, \varrho(t)\, s^z] + [s^z \varrho(t), s^z]\} . \end{aligned} \tag{3c.1}$$

The spin operators obey the commutation relations

$$\begin{aligned} &[s^z, s^\pm] = \pm s^\pm , \qquad [s^+, s^-] = 2s^z \\ &\quad (s^+)^2 = (s^-)^2 = 0 \quad (s^z)^2 = \tfrac{1}{4} . \end{aligned} \tag{3c.2}$$

The physical meaning of the transition rates γ_{10}, γ_{01}, and η becomes clear when equations of motion for the expectation values $\langle s^\pm \rangle$ and $\langle s^z \rangle$ are extracted from (3c.1). They are related to the transverse and longitudinal decay times T_2 and T_1, respectively and the equilibrium z-component of the spin by

$$\begin{aligned} T_2^{-1} &= \gamma_\perp = \tfrac{1}{2}(\gamma_{10}+\gamma_{01}+\eta) \\ T_1^{-1} &= \gamma_\parallel = \gamma_{10}+\gamma_{01} \\ \sigma_0 &= 2\langle s^z(t\to\infty)\rangle = (\gamma_{10}+\gamma_{01})/(\gamma_{01}-\gamma_{10}), \quad -1 \leqq \sigma_0 \leqq +1 . \end{aligned} \tag{3c.3}$$

Because of the wellknown analogy of spin-$\frac{1}{2}$ systems and two-level atoms the master equation (3c.1) also describes the behavior of a two-level atom under the influence of a heat bath. Then the operator $2s^z$ measures the population difference for exited state and ground state, whereas s^+ and s^- are the raising and lowering operators, respectively. Eq. (3c.1) may also be fancied up to describe spontaneous emission of electromagnetic radiation by an initially excited two-level atom [31]. In this case $\mathfrak{B}$ is the quantized electromagnetic field into which the atom dissipates its excitation energy.

4. Superconductors

4a) Introductory Remarks

The present-day understanding of superconductivity was initiated by two ideas. First, there was Fröhlich's suggestion to hold the interaction of conduction electrons and lattice vibrations responsible for the properties of superconducting metals [32]. The validity of this point was clearly demonstrated by the discovery of the isotope effect [33, 34]. Then Cooper [35] realized that the Fermi sea which is the ground state for free electrons is unstable with respect to formation of bound electron pairs, if there is an attractive interaction between the electrons. The BCS-theory [36] synthesized the two hints. Bardeen, Cooper and Schrieffer showed that the electron-phonon interaction in a metal can indeed produce an effective electron-electron attraction for electrons with energies E in the interval $E_F - \hbar\omega_D \lesssim E \lesssim E_F + \hbar\omega_D$ where E_F and ω_D are the Fermi energy and the Debye frequency, respectively. As a consequence, the superconducting ground state is a pair condensate with respect to these electrons. The lowest excited states, corresponding to quasi-particles and quasi-holes in the modified Fermi sea, were then found to be separated from the ground state by a finite energy gap Δ. The BCS-theory has since proved to give an at least qualitatively satisfying account of the thermodynamic, electromagnetic, and transport properties of most superconducting materials [37, 38]. Many of the quantitative discrepancies between theory and experiment, as found especially for the socalled strong-coupling superconductors like Pb and Hg have been eliminated by more sophisticated versions of the original BCS-theory [39]. Such sophistications mainly consist in a more detailed treatment of the electron-phonon interaction which Bardeen, Cooper and Schrieffer accounted for in terms of two parameters only, the Debye frequency ω_D and a coupling constant measuring the strength of the effective electron-electron attraction. The ensuing generalizations of the BCS-theory, obtained by Eliashberg [40] and Scalapino et al. [41] can be characterized as follows. First, the effective attraction between electrons produced by exchange of virtual phonons is a retarded interaction rather than an instantaneous one. As a consequence, the gap-parameter Δ assumes a time dependence or, equivalently, a frequency dependence, $\Delta \to \Delta(\omega)$. This frequency dependence is closely related to the phonon spectrum. Second, real phonons can be created and annihilated in the course of electron collisions. This effect causes the elementary excitations in the superconductor to have finite life times. Formally, the gap parameter becomes a complex number, $\Delta(\omega) = \Delta_1(\omega) + i\Delta_2(\omega)$. By measuring the tunnel current from a superconductor to a normal conductor through a thin insolating layer the complex frequency-dependant gap parameter $\Delta(\omega)$ can be determined experimentally. Experiments on the strong-coupling superconductors Pb and Hg have satisfyingly confirmed the predictions of Scalapino et al. [41].

The following treatment of superconductivity does not present anything physically new. It is meant as a demonstration of the applicability of generalized-master-equation techniques to nontrivial many-body problems. For this purpose it may suffice to consider a somewhat simplified model Hamiltonian for the interacting electrons ($\mathfrak{S}$) and phonons ($\mathfrak{B}$)

$$\begin{aligned} H_{\mathfrak{S}} \equiv H_{\mathrm{el}} &= \sum_{k\sigma} \varepsilon_k c^+_{k\sigma} c_{k\sigma} \\ H_{\mathfrak{B}} \equiv H_{\mathrm{ph}} &= \sum_{q} \omega_q b^\dagger_q b_q \\ H_{\mathfrak{S}\mathfrak{B}} \equiv H_{\mathrm{el\text{-}ph}} &= \sum_{kq\sigma} g_q c^+_{k+q,\sigma} c_{k\sigma} (b_q + b^\dagger_{-q}) \\ &\equiv \sum_{kq\sigma} c^+_{k+q,\sigma} c_{k\sigma} \varphi_q \,. \end{aligned} \tag{4a.1}$$

Here $c^+_{k\sigma}$ and $c_{k\sigma}$ create and annihilate, respectively, electrons with spin σ, wave vector $\boldsymbol{k}$, and energy ε_k. The electron energies ε_k are measured from the unperturbed Fermi level. The electron operators obey Fermi commutation relations. The (Bose) operators $b^\dagger_q$ and b_q create and annihilate, respectively, phonons with wave vector $\boldsymbol{q}$ and energy ω_q. The interaction is characterized by a coupling constant g_q. A realistic theory would have to account for the Coulomb interaction in $H_{\mathfrak{S}}$, more than one phonon branch in $H_{\mathfrak{B}}$, and a more general coupling $g_{k,q}$.

We will calculate the following (retarded) Green's functions

$$\begin{aligned} G(\boldsymbol{k}t) &= i\,\Theta(t)\,\langle[c_{\boldsymbol{k}\sigma}(t), c^+_{\boldsymbol{k}\sigma}(0)]_+\rangle \\ F^+(\boldsymbol{k}t) &= i\,\Theta(t)\,\langle[c^+_{-\boldsymbol{k},-\sigma}(t), c^+_{\boldsymbol{k}\sigma}(0)]_+\rangle \end{aligned} \tag{4a.2}$$

where $[\ldots,\ldots]_+$ denotes an anticommutator and

$$\langle\ldots\rangle = \mathrm{tr}_{\mathrm{el}}\,\mathrm{tr}_{\mathrm{ph}}\ldots e^{-\beta H}/\mathrm{tr}_{\mathrm{el}}\,\mathrm{tr}_{\mathrm{ph}}\,e^{-\beta H}\,.$$

The one-particle Green's function $G(\boldsymbol{k}t)$ contains all information about the behavior of single electrons, that is, e.g., the quasiparticle excitation spectrum. The anomalous Green's function or pair amplitude $F^+(\boldsymbol{k}t)$ vanishes identically for normal electron systems because of electron number conservation. Its nonvanishing for superconductors siquals that number conservation is broken because of pair condensation. According to Yang [42] $F^+ \neq 0$ implies off-diagonal long range order in the electronic system.

The Heisenberg equations of motion for electron and phonon operators imply equations of motion for $G(\boldsymbol{k}t)$, $F^+(\boldsymbol{k}t)$ and mixed electron-phonon Green's functions of more complicated structure. If the hierarchy of equations for all these Green's functions is suitably organized by using a method developed by Martin and Schwinger [43], a systematic perturbation expansion for $G(\boldsymbol{k}t)$ and $F^+(\boldsymbol{k}t)$ can be generated. This method was used in [40, 41]. Our procedure will be based on first eliminating the phonon degrees of freedom and then consider Nakajima-Zwanzig-type equations of motion for electron Green's functions.

4b) The Electrons as an Open System

In order to generate equations of motion for the Green's functions $G(\boldsymbol{k}t)$ and $F^+(\boldsymbol{k}t)$ we express these quantities in the form (2e.12) and (2e.13)

$$\begin{aligned} G(S_1, S_2, t) &= i\,\Theta(t)\,\langle[S_2(t), S_1(0)]_+\rangle \\ &= i\,\Theta(t)\,\mathrm{tr}_{\mathrm{el}}\,S_2\,V(t)\,[S_1, \varrho^{(0)}]_+ \\ &\quad + \Theta(t)\int_0^t \mathrm{d}t'\,\mathrm{tr}_{\mathrm{el}}\,S_2\,V(t-t')\,\mathrm{tr}_{\mathrm{ph}}\,L_{\mathrm{el\text{-}ph}}\,e^{-i(1-\mathfrak{P})Lt'} \\ &\quad \cdot (1-\mathfrak{P})\,[S_1, W(0)]_+\,. \end{aligned} \tag{4b.1}$$

Since we are considering a thermal equilibrium problem the initial density operator $W(0)$ of the composite system $\mathfrak{S}$ (electrons) $\oplus\, \mathfrak{B}$ (phonons) is taken as the canonical operator

$$W(0)=e^{-\beta H}/\mathrm{tr}_{el}\,\mathrm{tr}_{ph}\,e^{-\beta H}\,. \tag{4b.2}$$

The projection operator $\mathfrak{P}$ used to eliminate the phonon coordinates, is

$$\mathfrak{P}=B_{ref}\,\mathrm{tr}_{ph} \tag{4b.3}$$

with the free canonical operator

$$B_{ref}=e^{-\beta H_{ph}}/\mathrm{tr}_{ph}\,e^{-\beta H_{ph}} \tag{4b.4}$$

as the reference state for the phonons. The equation of motion (2b.17) for the time evolution operator $V(t)$ can be simplified a bit because of

$$\begin{aligned} &L_{ph}B_{ref}=\hbar^{-1}[H_{ph},B_{ref}]=0\\ &\mathrm{tr}_{ph}L_{el\text{-}ph}B_{ref}=0\,. \end{aligned} \tag{4b.5}$$

We then have

$$\begin{aligned} \dot V(t)=&-iL_{el}V(t)\\ &-\int_0^t dt'\,\mathrm{tr}_{ph}L_{el\text{-}ph}\,e^{-i(1-\mathfrak{P})Lt'}L_{el\text{-}ph}B_{ref}V(t-t')\,. \end{aligned} \tag{4b.6}$$

The integral kernel in this equation can be expanded in terms of the coupling constant $g_{\boldsymbol{q}}$ according to (2b.18). By an argument similar to Migdal's [44] it is easily shown that such an expansion actually goes in terms of the ratio $(m/M)^{1/2}$ of the electron mass and the mass of an ion in the lattice. Since $m/M \ll 1$ we can approximate the integral kernel in lowest order (Born approximation) to get

$$\begin{aligned} \dot V(t)=&-iL_{el}V(t)\\ &-\int_0^t dt'\,\mathrm{tr}_{ph}L_{el\text{-}ph}\,e^{-i(L_{el}+L_{ph})t'}L_{el\text{-}ph}B_{ref}V(t-t')\,. \end{aligned} \tag{4b.7}$$

By differentiating (4b.1) with respect to time and using (4a.1) we find, after some lengthy but trivial algebra, the following equations for $G(\boldsymbol{k}t)$ and $F^{+}(\boldsymbol{k}t)$

$$\begin{aligned} \left(\frac{\partial}{\partial t}+i\varepsilon_{\boldsymbol{k}}\right)G(\boldsymbol{k}t)-i\delta(t)=&\;i\int_0^t dt'\sum_{\boldsymbol{q}}\tfrac{1}{2}\{D^{(-)}(\boldsymbol{q}t)+D^{(+)}(\boldsymbol{q}t)\}\\ &\cdot e^{-i\varepsilon_{\boldsymbol{k}-\boldsymbol{q}}t}G(\boldsymbol{k},t-t')\\ &+i\int_0^t dt'\sum_{\boldsymbol{k}'\sigma'\boldsymbol{q}}D^{(-)}(\boldsymbol{q},t')\,e^{-i\varepsilon_{\boldsymbol{k}-\boldsymbol{q}}t'}\,i\Theta(t-t')\\ &\cdot\langle[c^{+}_{\boldsymbol{k}'-\boldsymbol{q},\sigma'}c_{\boldsymbol{k}'\sigma'}c_{\boldsymbol{k}-\boldsymbol{q},\sigma}(t-t'),c^{+}_{\boldsymbol{k}\sigma}(0)]_{+}\rangle \end{aligned}$$

and (4b.8)

$$\left(\frac{\partial}{\partial t}-i\varepsilon_{k}\right)F^{+}(k\,t)=-i\int_{0}^{t}\mathrm{d}t'\sum_{q}\tfrac{1}{2}\{D^{(-)}(q\,t)-D^{(+)}(q\,t)\}$$
$$\cdot\,\mathrm{e}^{+i\varepsilon_{k-q}t}F^{+}(k,t)$$
$$-i\int_{0}^{t}\mathrm{d}t'\sum_{k'\sigma' q}D^{(-)}(q,t')\,\mathrm{e}^{+i\varepsilon_{k-q}t'}\,i\Theta(t-t')$$
$$\cdot\langle[c^{+}_{-k+q,-\sigma}c^{+}_{k'-q,\sigma'}c_{k'\sigma'}(t-t'),c^{+}_{k\sigma}(0)]_{+}\rangle\,.$$

The influence of the phonons on the electrons expresses itself here in terms of the unperturbed equilibrium phonon Green's functions

$$D^{(\pm)}(q\,t)=i\Theta(t)\,\mathrm{tr}_{\mathrm{ph}}[\varphi_{q}(t),\varphi_{-q}(0)]_{\pm}B_{\mathrm{ref}} \tag{4b.9}$$

with

$$\varphi_{q}(t)=\mathrm{e}^{iH_{\mathrm{ph}}t/\hbar}\,\varphi_{q}(0)\,\mathrm{e}^{-iH_{\mathrm{ph}}t/\hbar}\,.$$

$D^{(+)}$ and $D^{(-)}$ are related by the fluctuation-dissipation theorem [26]. The further treatment of Eq. (4b.8) follows standard lines. First, the two-particle Green's functions occurring on the right hand sides are decoupled by a mean field ansatz

$$i\Theta(t)\,\langle[c^{+}_{k'-q,\sigma'}c_{k'\sigma'}c_{k-q,\sigma}(t),c^{+}_{k\sigma}(0)]_{+}\rangle$$
$$=\{\delta_{q,0}n(k')-\delta_{kk'}\delta_{\sigma\sigma'}n(k-q)\}\;G(k\,t) \tag{4b.10}$$
$$+\delta_{k',-k+q}\delta_{\sigma,-\sigma}f(k-q)\,F^{+}(k\,t)$$

and

$$i\Theta(t)\,\langle[c^{+}_{-k+q,-\sigma}c^{+}_{k'-q,\sigma}c_{k'\sigma'}(t),c^{+}_{k\sigma}(0)]_{+}\rangle$$
$$=\{\delta_{q,0}n(k')-\delta_{k',-k+q}\delta_{\sigma',-\sigma}n(k-q)\}\;F^{+}(k\,t)$$
$$+\delta_{k'k}\delta_{\sigma'\sigma}G(k\,t)\,.$$

This generalizes the Hartree-Fock approximation to include the pair amplitude $F^{+}(k\,t)$. The equilibrium expectation values $n(k)$ and $f(k)$ appearing in (4b.10) are related to the Fourier transforms of $G(k\,t)$ and $F^{+}(k\,t)$ by wellknown spectral theorems [26]

$$\begin{aligned} n(k)=\langle c^{+}_{k\sigma}c_{k\sigma}\rangle &= \int_{-\infty}^{+\infty}\frac{2\mathrm{d}\omega}{1+e^{\beta\omega}}\,\mathrm{Im}\,G(k,E)\Big|_{E=\omega+i0}\\ f(k)=\langle c^{+}_{-k,-\sigma}c^{+}_{k\sigma}\rangle &= \int_{-\infty}^{+\infty}\frac{2\mathrm{d}\omega}{1+e^{\beta\omega}}\,\mathrm{Im}\,F^{+}(k,E)\Big|_{E=\omega+i0}\,. \end{aligned} \tag{4b.11}$$

We will not discuss the limits of validity of the mean field approximation (4b.10). We should mention, however, that for bulk superconductors there is no experimental indication for the mean field theory to break down [45]. Moreover, Boguolyubov [46] has shown that for a slightly simplified model system the decoupling (4b.10) becomes exact in the thermodynamic limit. By inserting (4b.10) in (4b.9) and Fourier transforming with respect to time we get the following generalizations of Gorkov's equations

$$\begin{aligned}&[E-\sigma_{\boldsymbol{k}}+Z_-(\boldsymbol{k},E)]\,G(\boldsymbol{k},E)-\Phi_-(\boldsymbol{k},E)\,F^+(\boldsymbol{k},E)=-(1/2\pi)\\&-\Phi_+(\boldsymbol{k},E)\,G(\boldsymbol{k},E)+[E+\varepsilon_{\boldsymbol{k}}-Z_+(\boldsymbol{k},E)]\,F^+(\boldsymbol{k}E)=0\end{aligned}\tag{4b.12}$$

with

$$\begin{aligned}\Phi_\pm(\boldsymbol{k},E)&=\sum_{\boldsymbol{k}'}D^{(-)}(\boldsymbol{k}-\boldsymbol{k}',E\pm\varepsilon_{\boldsymbol{k}'})\,f(\boldsymbol{k}')\\Z_\pm(\boldsymbol{k},E)&=\sum_{\boldsymbol{k}'}[\tfrac{1}{2}\{D^{(-)}(\boldsymbol{k}-\boldsymbol{k}',E\pm\varepsilon_{\boldsymbol{k}'})\mp D^{(+)}(\boldsymbol{k}-\boldsymbol{k}',E\pm\varepsilon_{\boldsymbol{k}'})\}\\&\qquad-D^{(-)}(\boldsymbol{k}-\boldsymbol{k}',E\pm\varepsilon_{\boldsymbol{k}'})\,n(\boldsymbol{k}')]\,.\end{aligned}\tag{4b.13}$$

For the four renormalization amplitudes we find four coupled nonlinear integral equations by inserting in (4b.13) the spectral theorems (4b.11):

$$\begin{aligned}\Phi_\pm(\boldsymbol{k},E)&=-(2\pi)^{-1}\sum_{\boldsymbol{k}'}D^{(-)}(\boldsymbol{k}-\boldsymbol{k}',E\pm\varepsilon_{\boldsymbol{k}'})\\&\cdot\int_{-\infty}^{+\infty}2(1+\mathrm{e}^{-\beta\omega})\,\mathrm{d}\omega\cdot\mathrm{Im}\cdot\\&\frac{\Phi_+(\boldsymbol{k}',\omega+i0)}{[\omega-\varepsilon_{\boldsymbol{k}'}+Z_-(\boldsymbol{k}',\omega+i0)][\omega+\varepsilon_{\boldsymbol{k}'}-Z_+(\boldsymbol{k}',\omega+i0)]-\Phi_+(\boldsymbol{k}',\omega+i0)\Phi_-(\boldsymbol{k}',\omega+i0)}\end{aligned}\tag{4b.14}$$

and

$$\begin{aligned}Z_\pm(\boldsymbol{k},E)&=\tfrac{1}{2}\sum_{\boldsymbol{k}'}\{D^{(-)}(\boldsymbol{k}-\boldsymbol{k}',E\pm\varepsilon_{\boldsymbol{k}'})\mp D^{(+)}(\boldsymbol{k}-\boldsymbol{k}',E\pm\varepsilon_{\boldsymbol{k}'})\}\\&+(2\pi)^{-1}\sum_{\boldsymbol{k}'}D^{(-)}(\boldsymbol{k}-\boldsymbol{k}',E\pm\varepsilon_{\boldsymbol{k}'})\int_{-\infty}^{+\infty}2(1+\mathrm{e}^{\beta\omega})^{-1}\mathrm{d}\omega\cdot\mathrm{Im}\cdot\\&\frac{\omega+\varepsilon_{\boldsymbol{k}'}-Z_+(\boldsymbol{k}',\omega+i0)}{[\omega-\varepsilon_{\boldsymbol{k}'}+Z_-(\boldsymbol{k}',\omega+i0)][\omega+\varepsilon_{\boldsymbol{k}'}-Z_+(\boldsymbol{k}',\omega+i0)]-\Phi_+(\boldsymbol{k}',\omega+i0)\Phi_-(\boldsymbol{k}',\omega+i0)}\,.\end{aligned}\tag{4b.15}$$

We now explicitely see the generalizations obtained with respect to the BCS-theory. If we suppress the renormalization of the single electron energies $\varepsilon_{\boldsymbol{k}}$, i.e. put $Z_\pm=0$ and replace the phonon Green's function $D^{(-)}(\boldsymbol{k}-\boldsymbol{k}',E\pm\varepsilon_{\boldsymbol{k}'})$ in (4b.14) by an effective coupling constant $V_{\boldsymbol{k}\boldsymbol{k}'}$,

Eqs. (4b.12–15) reduce to the BCS-theory:

$$
\begin{aligned}
&\Phi_{\pm}(\boldsymbol{k},E)\to\Delta_{\boldsymbol{k}}=\sum_{\boldsymbol{k}'}V_{\boldsymbol{kk}'}f(\boldsymbol{k}')\\
&\quad G(\boldsymbol{k},E)\to-(2\pi)^{-1}\frac{E+\varepsilon_{\boldsymbol{k}}}{E^2-(\varepsilon_{\boldsymbol{k}}^2+\Delta_{\boldsymbol{k}}^2)}\\
&\quad F^{+}(\boldsymbol{k},E)\to-(2\pi)^{-1}\frac{\Delta_{\boldsymbol{k}}}{E^2-(\varepsilon_{\boldsymbol{k}}^2+\Delta_{\boldsymbol{k}}^2)}\\
&\Delta_{\boldsymbol{k}}=-(2\pi)^{-1}\sum_{\boldsymbol{k}'}V_{\boldsymbol{kk}'}\int_{-\infty}^{+\infty}2(1+\mathrm{e}^{-\beta\omega})^{-1}\,\mathrm{d}\omega\,\mathrm{Im}\frac{\Delta_{\boldsymbol{k}'}}{(\omega+i0)^2-(\varepsilon_{\boldsymbol{k}'}^2+\Delta_{\boldsymbol{k}'}^2)}\,.
\end{aligned}
\tag{4b.16}
$$

The selfconsistency equations (4b.14) and (4b.15) which correspond to the BCS gap equation are the same as those found in [41] except for the fact that the more general and more realistic Hamiltonian considered there entails some additional terms to appear in the selfconsistency equations.

5. Superradiance

5a) Introductory Remarks

As has been known since the beginnings of quantum theory, spontaneous emission of light is a quantum effect unexplicable in terms of classical physics. Nonetheless, the classical picture of emitters radiating in phase with each other can, under certain conditions, be used to understand the properties of light pulses spontaneously generated by a system of many excited atoms. If, for instance, N identical free atoms are prepared, at some instant of time, in an excited state of energy $\hbar\omega$ and if these atoms occupy a volume with linear dimensions $l \ll \lambda = 2\pi c/\omega$, then spontaneous emission generates a light pulse with mean intensity proportional to N^2. By energy conservation the spectral width of such a pulse is larger by a factor of the order N than the natural linewidth observed for independently radiating atoms. This effect was first discussed and termed superradiance by Dicke [47]. Dicke's 1954 paper has posed and left open a number of questions only recently answered by several authors [48–60]. One of these open questions was whether superradiance would ever be observable since the condition $l \ll \lambda$ precludes getting a sizable number of atoms involved. We will show here that superradiance can be produced under much weaker and in fact realizable conditions [53]. In order to get a quick survey over the physics of the problem we will first present a semiclassical discussion before entering the fully quantum-mechanical treatment.

5b) Semiclassical Theory

The propagation of a light pulse with carrier frequency ω through a medium of identical two-level atoms with transition frequency ω is semiclassically described by Maxwell's equations for the electric field $\boldsymbol{E}(x,t)$ and quantum-mechanical equations of motion for the spatial densities of the atomic variables polarization $\boldsymbol{P}(x,t)$ and inversion $D(x,t)$.

Our interest is in the special case with respect to the duration τ of the pulse

$$\omega^{-1} \ll \tau \ll T_1, T_2\,, \tag{5b.1}$$

where T_1 and T_2 are the relaxation times of inversion and polarization, respectively. The left hand condition ensures that a carrier frequency of the pulse can be defined. We have to pose the right hand condition, since a superradiant pulse can be generated by in-phase cooperation of all atoms only and since phase correlations between the atoms cannot persist for times larger than the relaxation times T_1 and T_2. In the special case (5b.1) we may use slowly varying field variables

$$\begin{aligned}
&\boldsymbol{E}(x,t) = \boldsymbol{E}^*(x,t) = \boldsymbol{e}\, i\sqrt{2\pi\hbar\,\omega/V}\,\{b(x,t)\,\mathrm{e}^{-i(\omega t-kx)} - b^\dagger(x,t)\,\mathrm{e}^{+i(\omega t-kx)}\}\\
&\boldsymbol{P}(x,t) = \boldsymbol{P}^*(x,t) = -\boldsymbol{e}\, i(\mu/V)\,\{S^-(x,t)\,\mathrm{e}^{-i(\omega t-kx)} - S^+(x,t)\,\mathrm{e}^{+i(\omega t-kx)}\}\\
&D(x,t) = D^*(x,t) = V^{-1}\,2S^z(x,t)
\end{aligned} \tag{5b.2}$$

with

$$\begin{Bmatrix}\partial/\partial t\\ \partial/\partial x\end{Bmatrix}\{b, b^\dagger, S^+, S^-, S^z\} \ll \begin{Bmatrix}\omega\\ \omega/c\end{Bmatrix}\{b, b^\dagger, S^+, S^-, S^z\}\,. \tag{5b.3}$$

There we have assumed the pulse to move into the positive x-direction and that all quantities depend on one spatial variable x only. This assumption will be justified later. Moreover, for simplicity we assume linear polarization as indicated by the unit vector $\boldsymbol{e}$. μ is the component of the atomic dipole moment in the direction $\boldsymbol{e}$. The normalization of the dimensionless variables b, $b^\dagger$, and S^α has been fixed in anticipation of the quantum-mechanical meaning these quantities we will take on in the next subsection. Since the volume V contains N atoms we have $|S^z|$, $|S^\pm| \leqq N/2$. The wellknown equations of motion mentioned above now read [61–63]

$$\begin{aligned}
&\left(\frac{\partial}{\partial t} + c\frac{\partial}{\partial x} + \kappa\right) b^\dagger(x,t) = i g\, S^+(x,t)\\
&\qquad\frac{\partial}{\partial t} S^+(x,t) = -i2g\, b^\dagger(x,t)\, S^z(x,t)\\
&\qquad\frac{\partial}{\partial t} S^z(x,t) = i g\{b^\dagger(x,t)\, S^-(x,t) - b(x,t)\, S^+(x,t)\}
\end{aligned} \tag{5b.4}$$

with the coupling constant

$$g=\mu\sqrt{2\pi\omega/\hbar V}\,. \tag{5b.5}$$

For the sake of generality we have accounted for losses of the electromagnetic field measured by the field damping constant κ. We may solve Eq. (5b.4) in terms of the ansatz

$$\begin{aligned} S^-(x,t)=S^+(x,t)&=\tfrac{1}{2}N\sin\Phi(x,t)\\ S^z(x,t)&=\tfrac{1}{2}N\cos\Phi(x,t)\\ b^\dagger(x,t)=-b(x,t)&=(i/2g)\,\partial\Phi(x,t)/\partial t\,. \end{aligned} \tag{5b.6}$$

The quantity $\Phi(x,t)$ is usually called Bloch angle. It characterizes the state of the $N\,\mathrm{d}V/V$ atoms in a volume element $\mathrm{d}V$ at the point x and at time t. It obeys the equation [64]

$$\left(\frac{\partial^2}{\partial t^2}+c\frac{\partial^2}{\partial x\partial t}+\kappa\frac{\partial}{\partial t}\right)\Phi(x,t)=Ng^2\sin\Phi(x,t)\,. \tag{5b.7}$$

The coefficient Ng^2 appearing here can be expressed in terms of measurable quantities as

$$Ng^2=c^2/l_c^2\,,\quad l_c=\sqrt{2\pi c/\varrho\gamma\lambda^2}\,,\quad \varrho=N/V\,,\quad \gamma=(8\pi^2/3)\,(\mu^2/\hbar\lambda^3)\,, \tag{5b.8}$$

where γ is the natural linewidth of the atomic transition. The material constant l_c is called cooperation length [64]. Its physical meaning is obvious from Eq. (5b.7). It gives the scale of length on which $\Phi(x,t)$ changes spatially due to the atom-field interaction. In order to describe the superradiant behavior of the atomic medium we have to solve Eq. (5b.7) with a suitable boundary condition at the end faces $x=0$, $x=l$ and with an appropriate initial condition $\Phi(x,0)=\Phi_0(x)$. The simplest initial condition is

$$\Phi(x,0)=\Phi_0=\text{const}\quad\text{in}\quad 0\leqq x\leqq l\,. \tag{5b.9}$$

Let us quickly ascertain under which conditions such an initial state can be prepared experimentally. The common preparation technique consists in first bringing all atoms to the ground state and then sending a laser pulse along the axis of the sample. The (constant) amplitude b_0 and the duration T of this pump pulse are chosen such that $T\ll l/c$ and $\Phi_0=|2gb_0T|$. The propagation of this pulse through the sample can be described by Eq. (5b.7) with, in general, $\kappa=0$. At time $t=l/c$ after the penetration of the pulse the sample is left in a state characterized by the Bloch angle $\Phi_0(x)$. This can be independent of x only if

$$l\ll l_c \tag{5b.10}$$

since the spatial variation of $\Phi_0(x)$ is on a scale l_c[5]. We require (5b.10) and now use Eq. (5b.7) and the initial condition (5b.9) to study superradiant pulses. We specify the damping constant κ as

$$\kappa = c/l\,. \tag{5b.11}$$

This damping is meant to simulate the "losses" of field energy due to the escape of light through the end faces of the sample. Note that l/c is just the time of flight of a photon through the sample [53]. This schematic way of dealing with leakage effects, which we have taken over from laser theory [61], saves us from having to account for complicated boundary conditions at $x=0$ and $x=l$. By using (5b.11) we may state the limit (5b.10) in the alternative form

$$g\sqrt{N}/\kappa = l/l_c \ll 1\,. \tag{5b.12}$$

In this limit the Bloch angle $\Phi(x,t)$ can be determined quite easily. To obtain it we rewrite Eq. (5b.7) in terms of dimensionless variables

$$t/\tau = t g^2 N/\kappa\,, \quad x/\xi = x g^2 N/c\kappa \tag{5b.13}$$

as

$$\frac{\partial}{\partial(t/\tau)}\Phi - \sin\Phi = -(l^2/l_c^2)\left(\frac{\partial^2}{\partial(t/\tau)^2} + \frac{\partial^2}{\partial(t/\tau)\,\partial(x/\xi)}\right)\Phi\,. \tag{5b.14}$$

To lowest order in l^2/l_c^2 the Bloch angle Φ can thus be obtained from

$$\dot{\Phi} = \tau^{-1}\sin\Phi \tag{5b.15}$$

and will not depend on the spatial variable x. This approximate equation for Φ corresponds to an adiabatic elimination of the electric field b from (5b.6) by

$$b = -igS^-/\kappa\,. \tag{5b.16}$$

We recognize Eq. (5b.15) as the equation of motion for an overdamped pendulum. Its solution reads

$$\tanh\frac{\Phi(t)}{2} = e^{t/\tau}\tanh\frac{\Phi_0}{2}\,. \tag{5b.17}$$

This implies the following result for the radiated intensity

$$\begin{aligned}2\kappa\hbar\omega\, b^\dagger(t)\, b(t) &= \hbar\omega I_1 S^+(t)\, S^-(t)\\ &= \hbar\omega I_1 (N/2)^2\,\mathrm{sech}^2[(t-t_{\max})/\tau]\end{aligned} \tag{5b.18}$$

[5] Let us remark that a spatially homogeneous complete inversion of the atomic population ($\Phi_0 = 0$) can in principle be realized without any restriction on the length l of the sample, if an incoherent transverse excitation mechanism is used.

with $I_1 = 2g^2/\kappa$ and

$$\begin{aligned} t_{\max} &= -\tfrac{1}{2}\tau \ln \tanh(\Phi_0/2) \\ &= \tfrac{1}{2}\tau \ln[(\tfrac{1}{2}N + S^z(0))/(\tfrac{1}{2}N - S^z(0))]\,. \end{aligned} \tag{5b.19}$$

The maximum intensity, reached at time $t = t_{\max}$, is proportional to N^2 as is characteristic for a superradiant pulse. If all atoms are excited initially, i.e. $\Phi_0 = 0$ or $S^z(0) = +N/2$, then the semiclassical theory presented here gives the nonsensical result $t_{\max} = \infty$. The pendulum described by Eq. (5b.15) has an unstable equilibrium position at $\Phi = 0$. The semiclassical treatment can be valid only if the atoms display a non-zero initial polarization $S^{\pm}(0) = [N^2/4 - S^z(0)^2]^{1/2}$. The quantum mechanical theory to be given in the next subsection will not have this drawback and will reveal the precise limit of validity of the semiclassical theory with respect to the initial Bloch angle Φ_0. In order to find the physical meaning of the expressions $I_1 = 2g^2/\kappa$ and $\tau = \kappa/g^2 N$ and to justify our considering one spatial variable x only we specify the shape of the active volume as that of a long thin cylinder with length l and diameter d such that

$$\lambda \ll d \ll l\,. \tag{5b.20}$$

For this case it is intuitively clear and may be shown by a more detailed analysis [48, 52] that only that part of the radiation which goes into the small diffraction solid angle

$$\Delta\Omega = \lambda^2/d^2 \ll 1 \tag{5b.21}$$

around the axis of the cylinder is enhanced by cooperative effects. Up to corrections of order $1/N$ the whole initial atomic excitation energy is radiated into the superradiant pulse travelling along the axis of the sample. We then have

$$\begin{aligned} 1/\tau &= \gamma N \Delta\Omega/4\pi \\ \hbar\omega I_1 &= \hbar\omega\gamma\Delta\Omega/4\pi\,. \end{aligned} \tag{5b.22}$$

The linewidth $1/\tau$ of the superradiant pulse is thus seen to be the natural linewidth γ of the atomic transition enhanced by the factor N and reduced by the geometry factor $\Delta\Omega/4\pi$. Correspondingly, $\hbar\omega I_1$ is the fraction of the intensity of a single-atom emission going into the diffraction solid angle $\Delta\Omega$.

Let us conclude by assembling the conditions for superradiance to occur

$$\begin{aligned} &\lambda \ll d \ll l \\ &l_{\text{abs}} = 4\pi/T_2\varrho\gamma\lambda^2 \ll l \ll l_c = \sqrt{2\pi c/\varrho\gamma\lambda^2}\,. \end{aligned} \tag{5b.23}$$

l_{abs} is the wellknown absorption length occurring in Beer's law. The difference to the case considered by Dicke [47] is obvious. Here the linear dimensions of the active volume have to be large compared to the wavelength λ associated with the atomic transition. The requirement $l_{\text{abs}} \ll l$, following from (5b.1), ensures the atomic decay to be dominated by cooperative effects rather than by incoherent relaxation. The condition $l \ll l_c$ guarantees the pulse to be quasimonochromatic, i.e. propagation effects within the sample to be negligible. Let us note that the pulse duration τ cannot be made arbitrarily short by indefinitly increasing the number of atoms N. We rather have

$$1/\tau \approx \varrho \gamma \lambda^2 l \approx c l/l_c^2 \ll c/l \ll c/\lambda \approx \omega . \tag{5b.24}$$

5c) Superradiance Master Equation

We consider the system specified by the conditions (5b.22). The complex amplitudes b and $b^\dagger$ of the quasimonochromatic light field and the atomic variables S^α are now operators obeying the commutation rules

$$\begin{aligned} &[b, b^\dagger] = 1 , \quad [b, b] = [b^\dagger, b^\dagger] = 0 \\ &[S^z, S^\pm] = \pm S^\pm , \quad [S^+, S^-] = 2S^z . \end{aligned} \tag{5c.1}$$

The dynamics of the system is described by the equation of motion for the density operator $W(t)$ of the radiating atoms and the light field

$$\dot{W}(t) = -iLW(t) , \quad L = L_A + L_F + L_{AF} + i\Lambda_F . \tag{5c.2}$$

The three reversible parts of the Liouvillian L denote commutators with the Hamiltonians

$$\begin{aligned} H_A &= \hbar \omega S^z \\ H_F &= \hbar \omega b^\dagger b \\ H_{AF} &= \hbar g (b S^+ + b^+ S^-) , \end{aligned} \tag{5c.3}$$

where g is the coupling constant given in (5b.5). The irreversible part Λ_F of the Liouvillian represents a field damping accounting for leakage of photons through the end faces of the active volume. It acts on some operator X as

$$\Lambda_F X = \kappa \{ [b, X b^\dagger] + [b X, b^\dagger] \} . \tag{5c.4}$$

We recognize Λ_F as the damping Liouvillian for a damped harmonic oscillator moving under the influence of a zero temperature reservoir[6]. For later use we note that (5c.4) implies

$$\operatorname{tr} b^{\dagger l} b^{l'} e^{\Lambda_F t} X = e^{-(l+l')\kappa t} \operatorname{tr} b^{\dagger l} b^{l'} X . \tag{5c.5}$$

[6] Compare (3b.7) and (3b.9).

This is easily proved by differentiating with respect to time and using the commutation rules (5c.1). In the interaction picture Eq. (5c.2) simplifies, because of the atoms being in resonance with the field mode, to

$$\dot{\tilde{W}}(t) = -i(L_{AF} + i\Lambda_F)\,\tilde{W}(t)\,. \tag{5c.6}$$

We will in the following exclusively work in the interaction picture and may thus drop the tilde, $\tilde{W} \to W$. The atom-field interaction and the field damping are characterized, as in the semiclassical theory, by the time scales $1/g\sqrt{N}$ and $1/\kappa$ respectively. The order-of-magnitude relation (5b.12) may be written as

$$O(L_{AF}) = g\sqrt{N}\,,\quad O(\Lambda_F) = \kappa\,,\quad O(L_{AF})/O(\Lambda_F) = \frac{g\sqrt{N}}{\kappa} = \frac{l}{l_c} \ll 1\,. \tag{5c.7}$$

The order-of-magnitude estimate for the field damping Liouvillian, $O(\Lambda_F) = \kappa$, can be read off the definition (5c.5). On the other hand, the estimate $O(L_{AF}) = g\sqrt{N}$ cannot be gained from just looking at H_{AF}; it is intuitive, though, in view of the semiclassical treatment in the foregoing subsection and can, moreover, be obtained by an analysis of the eigenvalue spectrum of H_{AF} [65, 66]. Scharf [65] has found that for $N \gg 1$ the eigenvalues of H_{AF} are nearly equidistant and separated by $\sim \hbar g\sqrt{N}$. As a consequence, expectation values of observables of the system will in general display a quasiperiodicity with quasiperiod $g\sqrt{N}$, if the state of the system is a superposition or mixture of many eigenstates of H_{AF}. The order-of-magnitude relation (5c.7) implies, as in the semiclassical theory that the field will follow the motion of the atoms adiabatically.

Since we are interested in a spontaneous emission effect we want to solve Eq. (5c.6) with the initial condition

$$W(0) = |0\rangle\langle 0| \otimes \varrho(0)\,, \tag{5c.8}$$

where $|0\rangle$ is the photon vacuum, $b|0\rangle = 0$, and $\varrho(0)$ the initial value of the atomic density operator

$$\varrho(t) = \mathrm{tr}_F\, W(t)\,. \tag{5c.9}$$

We think of the atomic initial state as being prepared by the method discussed in the preceding subsection, that is by means of an intense short laser pulse. If this pump pulse has a random phase and a fixed stable amplitude, it generates an atomic state which has been shown by Bonifacio, Haake, and Schwendimann [54] to be well approximated by

$$\varrho(0) = |\tfrac{1}{2}N, m\rangle\langle\tfrac{1}{2}N, m|$$

with

$$\begin{aligned} S^z|\tfrac{1}{2}N, m\rangle &= m|\tfrac{1}{2}N, m\rangle \\ S^\pm|\tfrac{1}{2}N, m\rangle &= [(\tfrac{1}{2}N \mp m)(\tfrac{1}{2}N \pm m + 1)]^{1/2}|\tfrac{1}{2}N, m \pm 1\rangle\,. \end{aligned} \tag{5c.10}$$

The "angular momentum" quantum number m is related to the Bloch angle Φ by $m=\frac{1}{2}N\cos\Phi$ where $\frac{1}{2}N\cos\Phi$ has to be rounded to the next-lying integer.

In view of (5c.7) it is advantageous to adiabatically eliminate the field coordinates from Eq. (5c.6) and to study the Nakajima-Zwanzig equation for the atomic density operator $\varrho(t)$. To this end we define the projector

$$\mathfrak{P}=|0\rangle\langle 0|\,\mathrm{tr}_F\,. \tag{5c.11}$$

Here we have chosen the initially present photon vacuum as the reference state for the field. Because of

$$\mathfrak{P}L_{AF}\mathfrak{P}=0\,,\quad \Lambda_F|0\rangle\langle 0|=0\,,\quad (1-\mathfrak{P})\,W(0)=0\,,\quad \mathfrak{P}\Lambda_F=0 \tag{5c.12}$$

the Nakajima-Zwanzig equation here reads

$$\dot{\varrho}(t)=-\int_0^t \mathrm{d}t'\,\mathrm{tr}_F L_{AF}\,\mathrm{e}^{-i[i\Lambda_F+(1-\mathfrak{P})L_{AF}]t'}L_{AF}|0\rangle\langle 0|\varrho(t-t')\,. \tag{5c.13}$$

Up to corrections of higher order in l/l_c we can approximate the integral kernel in (5c.13) in lowest order in L_{AF}. By using (5c.5) we thus obtain

$$\dot{\varrho}(t)=\int_0^t \mathrm{d}t'\kappa\,\mathrm{e}^{-\kappa t'}\Lambda_c\varrho(t-t') \tag{5c.14}$$

with the collective decay Liouvillian

$$\Lambda_c X=\tfrac{1}{2}I_1\{[S^-,XS^+]+[S^-X,S^+]\}\,. \tag{5c.15}$$

The order of magnitude of Λ_c can be found as

$$O(\Lambda_c)=O\left(\int_0^\infty \mathrm{d}t\,L_{AF}\mathrm{e}^{\Lambda_F t}L_{AF}\right)=g^2N\int_0^\infty \mathrm{d}t\,\mathrm{e}^{-\kappa t}=g^2N/\kappa=1/\tau\,. \tag{5c.16}$$

This is just the inverse pulse duration known from the semiclassical treatment of the preceding subsection. Since we have $\kappa\gg 1/\tau$ we can neglect retardation effects in (5c.14) and so finally get the quantum mechanical analog of the semiclassical equation (5b.15) for the Bloch angle

$$\dot{\varrho}(t)=\Lambda_c\varrho(t)\,. \tag{5c.17}$$

Before proceeding to solve this superradiance master equation [53, 54] let us first construct the analog of Eq. (5b.16) in order to see explicitly how the field follows the motion of the atoms adiabatically. To this end we have to evaluate the expectation values $\langle b^\dagger(t)^l\, b(t)^{l'}\rangle$ which for $l, l'=0, 1, 2, \ldots$ specify the statistical properties of the radiation field.

We use (2b.12) and obtain

$$
\begin{aligned}
\langle b^\dagger(t)^l b(t)^{l'}\rangle &= \mathrm{tr}_A \mathrm{tr}_F b^{\dagger l} b^{l'} \{\mathfrak{P} W(t) + (1-\mathfrak{P}) W(t)\} \\
&= -i \int_0^t \mathrm{d}t' \mathrm{tr}_A \mathrm{tr}_F b^{\dagger l} b^{l'} e^{[\Lambda_F - i(1-\mathfrak{P})L_{AF}]t'} L_{AF}|0\rangle \langle 0|\varrho(t-t') \\
&= \sum_{n=0}^{\infty} (-i)^{n+1} \int_0^t \mathrm{d}t' \int_0^{t'} \mathrm{d}s_n \int_0^{s_n} \mathrm{d}s_{n-1} \dots \int_0^{s_2} \mathrm{d}s_1 \mathrm{tr}_A \mathrm{tr}_F b^{\dagger l} b^{l'} \\
&\quad \cdot U(t'-s_n) \mathfrak{Q} L_{AF} U(s_n - s_{n-1}) \mathfrak{Q} L_{AF} U(s_{n-1} - s_{n-2}) \\
&\quad \dots \mathfrak{Q} L_{AF} U(s_1) L_{AF}|0\rangle \langle 0|\varrho(t-t') .
\end{aligned}
\tag{5c.18}
$$

Here we have immediately expanded the exponential $\exp[\Lambda_F - i(1-\mathfrak{P}) L_{AF}]t$ in terms of L_{AF} and used the abbreviations $U(t) = \exp \Lambda_F t$ and $\mathfrak{Q} = 1 - \mathfrak{P}$. Because of $O(U(t)) = \exp(-\kappa t)$ and $O(L_{AF}) = g\sqrt{N}$ the above expansion goes in terms of the small parameter $g\sqrt{N}/\kappa$ and may thus be truncated after the first nontrivial term. The first nonvanishing term in the series arises for $n+1 = l+l'$. This can be seen as follows. Let us disregard, for the moment being, the factors $U\mathfrak{Q}$ which are irrelevant for the argument. By using the cyclic invariance of the trace we let all $n+1$ factors L_{AF} in the n-th term of the series act to the left. When $b^{\dagger l} b^{l'}$ is thus $(n+1)$ times commuted with H_{AF} a polynomial in b^+ and b is obtained. All monomials $b^{\dagger i} b^j$ in this polynomial are characterized by $(i+j) \geqq (l+l') - (n+1)$ because of the commutation relations $[b, f(b, b^\dagger)] = \partial f(b, b^\dagger)/\partial b^\dagger$. Now for $(n+1) < (l+l')$, i.e. $(i+j) > 0$, we have $\mathrm{tr}_F b^{\dagger i} b^j |0\rangle \langle 0| = 0$. We thus see that the term of order $n+1 = l+l'$ is the first to produce a contribution $b^{\dagger i} b^j$ with $i = j = 0$ whose vacuum expectation value does not vanish. By the same reasoning we find that in the term of order $n+1 = l+l'$ we can replace the factors $\mathfrak{Q} = (1-\mathfrak{P})$ with unity. The field expectation values $\langle b^+(t)^l b(t)^{l'}\rangle$ are now expressed as

$$
\begin{aligned}
\langle b^\dagger(t)^l b(t)^{l'}\rangle &= (-i)^{l+l'} \int_0^t \mathrm{d}t' \int_0^{t'} \mathrm{d}s_{l+l'-1} \int_0^{s_{l+l'-1}} \mathrm{d}s_{l+l'-2} \dots \int_0^{s_2} \mathrm{d}s_1 \\
&\quad \cdot \mathrm{tr}_A \mathrm{tr}_F b^{+l} b^{l'} U(t' - s_{l+l'-1}) L_{AF} U(s_{l+l'-1} - s_{l+l'-2}) L_{AF} \\
&\quad \dots L_{AF} U(s_2 - s_1) L_{AF} U(s_1) L_{AF}|0\rangle \langle 0|\varrho(t-t') .
\end{aligned}
\tag{5c.19}
$$

To evaluate this further we proceed step by step as follows. The first damping propagator, $U(t' - s_{l+l'-1})$ is replaced with $\exp[-\kappa(t' - s_{l+l'-1})]$ because of (5c.5). The following commutator is made to act to the left as $[b^{+l} b^{l'}, H_{AF}] = g\{-l S^+ b^{+l-1} b^{l'} + l' S^- b^{+l} b^{l'-1}\}$. Then the next damping propagator, $U(s_{l+l'-1} - s_{l+l'-2})$ gives way to $\exp[-\kappa(s_{l+l'-1}$

$-s_{l+l'-2})]$. Going on in this way we obtain

$$\langle b^\dagger(t)^l b(t)^{l'}\rangle = (-ig)^{l+l'}(-1)^l(l+l')!\int_0^t dt' e^{-(l+l')\kappa t'} \mathrm{tr}_A S^{+l} S^{-l'} \varrho(t-t') \cdot \int_0^{t'} ds_{l+l'-1} \int_0^{s_{l+l'-1}} ds_{l+l'-2} \dots \int_0^{s_2} ds_1 e^{+\kappa(s_{l+l'-1}+s_{l+l'-2}+\dots+s_2+s_1)}. \tag{5c.20}$$

The $(l+l'-1)$-fold integral over the s_i is elementary and leads to

$$\langle b^\dagger(t)^l b(t)^{l'}\rangle = (-ig)^{l+l'}(-1)^l(l+l')\int_0^t dt' e^{-(l+l')\kappa t'} \cdot (e^{\kappa t'}-1)^{l+l'-1} \langle S^+(t-t')^l S^-(t-t')^{l'}\rangle. \tag{5c.21}$$

For times large compared to the photon transient time, $t \gg \kappa^{-1}$, retardation effects can be neglected so that we finally get

$$\langle b^\dagger(t)^l b(t)^{l'}\rangle = \mu^{*l}\mu^{l'}\langle S^+(t)^l S^-(t)^{l'}\rangle, \quad \mu = -ig/\kappa = -il/l_c\sqrt{N}. \tag{5c.22}$$

This is the quantum mechanical version of (5b.16). Field expectation values can thus be evaluated as soon as the superradiance master equation (5c.17) for the atomic density operator is solved.

5d) Solution of the Superradiance Master Equation

α) Quasiprobability Distribution $P(s, s, s, t)$

The simplest and physically most transparent method of solving the master equation (5c.17) for the atomic density operator consists in first transforming this equation into a partial differential equation for the following quasiprobability distribution function [56, 58].

$$P(s, s^*, s^z, t) = \int_{\mathrm{Re}\,\xi=-\infty}^{+\infty} \int_{\mathrm{Im}\,\xi=-\infty}^{+\infty} \int_{\eta=-\infty}^{+\infty} (d^2\xi/\pi)(d\eta/2\pi) \mathrm{tr}_A e^{-i\xi^*(s^*-S^+)} e^{-i\eta(s^z-S^z)} e^{-i\xi(s-S^-)} \varrho(t). \tag{5d.1}$$

This definition associates the c-number variables s, s^*, s^z with the operators S^-, S^+, S^z such that expectation values of operators are given by moments of the quasiprobability distribution function P as

$$\langle S^+(t)^l S^z(t)^k S^-(t)^{l'}\rangle = \mathrm{tr}_A S^{+l} S^{zk} S^{-l'} \varrho(t) = \int_{\mathrm{Re}\,s=-\infty}^{+\infty} \int_{\mathrm{Im}\,s=-\infty}^{+\infty} \int_{s^z=-\infty}^{+\infty} d^2s\, ds^z s^{*l} s^{zk} s^{l'} P(s, s^*, s^z, t). \tag{5d.2}$$

$P(s, s^*, s^z, t)$ is real but not necessarily positive. For a discussion of the mathematical properties of quasiprobability distributions we refer to [9–14].

A similar quasiprobability can be defined for the radiation field

$$P_F(\beta, \beta^*, t) = \int_{\mathrm{Re}\,\xi=-\infty}^{+\infty} \int_{\mathrm{Im}\,\xi=-\infty}^{+\infty} (\mathrm{d}^2\xi/\pi)\, \mathrm{tr}_F\, \mathrm{e}^{-i\xi^*(\beta^*-b^+)}\, \mathrm{e}^{-i\xi(\beta-b)}\, \varrho_F(t), \quad (5\mathrm{d}.3)$$

where $\varrho_F(t)$ is the density operator for the field. Here c-number variables are associated with operators such that

$$\langle b^\dagger(t)^l\, b(t)^{l'} \rangle = \int \mathrm{d}^2\beta\, \beta^{*l} \beta^{l'}\, P_F(\beta, \beta^*, t)\,. \quad (5\mathrm{d}.4)$$

$P_F(\beta, \beta^*, t)$ is the weight function in the diagonal representation of $\varrho_F(t)$ with respect to coherent states introduced by Glauber [11] which we have already used in Section 3. In the present case $P_F(\beta, \beta^*, t)$ is related to the quasiprobability distribution function $P(s, s^*, s^z, t)$ for the atomic variables by

$$P_F(\beta, \beta^*, t) = \frac{1}{|\mu|^2} \int_{-\infty}^{+\infty} \mathrm{d}s^z\, P\left(\frac{1}{\mu}\beta, \frac{1}{\mu}, \beta^*, s^z, t\right). \quad (5\mathrm{d}.5)$$

This follows from the adiabatic correspondence law (5c.22).

β) Equation of Motion for $P(s, s^*, s^z, t)$

In view of a theorem proved by Haken [15] it must be possible to convert the superradiance master equation (5c.17) to a differential equation of motion for the quasiprobability $P(s, s^*, s^z, t)$ because the set of operators S^+, S^-, S^z is closed with respect to commutations. To construct this equation we differentiate (5d.1) with respect to time and insert $\dot{\varrho} = \Lambda_c \varrho$. Then we use the identities

$$\begin{aligned} S^-\, \mathrm{e}^{i\xi^* S^+} &= \mathrm{e}^{i\xi^* S^*}\{-(i\xi^*)^2 S^+ - 2i\xi^* S^z + S^-\} \\ \mathrm{e}^{i\xi S^-}\, S^+ &= \{-(i\xi)^2 S^- - 2i\xi S^z + S^+\}\, \mathrm{e}^{i\xi S^-} \\ S^-\, \mathrm{e}^{i\eta S^z} &= \mathrm{e}^{i\eta S^z}\, S^-\, \mathrm{e}^{i\eta} \\ \mathrm{e}^{i\eta S^z}\, S^+ &= S^+\, \mathrm{e}^{i\eta S^z}\, \mathrm{e}^{i\eta}\,, \end{aligned} \quad (5\mathrm{d}.6)$$

which follow from the commutation relations (5c.1) and the following differential relations

$$\begin{aligned} S^+\, \mathrm{e}^{i\xi^* S^+} &= \frac{\partial}{\partial i\xi^*}\, \mathrm{e}^{i\xi^* S^+} \\ S^-\, \mathrm{e}^{i\xi S^-} &= \frac{\partial}{\partial i\xi}\, \mathrm{e}^{i\xi S^-} \\ S^z\, \mathrm{e}^{i\eta S^z} &= \frac{\partial}{\partial i\eta}\, \mathrm{e}^{i\eta S^z}\,. \end{aligned} \quad (5\mathrm{d}.7)$$

We thus get

$$\dot{P}(s, s^*, s^z, t) = I_1 \left\{ -\frac{\partial}{\partial s^*} s^* s^z - \frac{\partial}{\partial s} s s^z + \frac{\partial}{\partial s^z} s s^* + \tfrac{1}{2}\left(\frac{\partial^2}{\partial s^{*2}} s^{*2} + \frac{\partial^2}{\partial s^2} s^2\right) + \left(1 - \frac{\partial}{\partial s^z} - e^{-\partial/\partial s^z}\right)\right\} P(s, s^*, s^z, t). \tag{5d.8}$$

We may estimate the relative weight of the various terms in (5d.8) by referring the variables t, s, s^*, s^z to their respective scales

$$t \to t/\tau \qquad \tau = \kappa/g^2 N = 2/I_1 N$$
$$s^\alpha \to s^\alpha/(\tfrac{1}{2}N) \quad \text{because of} \quad |\langle S^\alpha \rangle| \leqq N/2. \tag{5d.9}$$

In this way we see that j-th order derivatives in (5d.8) have the weight $(2/N)^{j-1}$. Up to corrections of order $1/N \ll 1$ the quasi-probability $P(s, s^*, s^z, t)$ therefore obeys the first order differential equation

$$\dot{P}(s, s^*, s^z, t) = I_1 \left\{ -\frac{\partial}{\partial s^*} s^* s^z - \frac{\partial}{\partial s} s s^z + \frac{\partial}{\partial s^z} s s^* \right\} P(s, s^*, s^z, t). \tag{5d.10}$$

This equation allows for solutions depending on s^z and the product $s s^*$ only. As we will see below the initial quasiprobabilities relevant for us obey

$$P(s, s^*, s^z, t=0) = P(s s^*, s^z, t=0). \tag{5d.11}$$

We can therefore determine the time-dependent quasiprobability from

$$\dot{P}(s s^*, s^z, t) = I_1 \left\{ -2 \frac{\partial}{\partial s s^*} s s^* s^z + \frac{\partial}{\partial s^z} s s^* \right\} P(s s^*, s^z, t). \tag{5d.12}$$

γ) Solution of the Initial Value Problem

The first-order differential equation (5d.12) can be solved using the method of characteristics. The associated characteristic curves are defined by the following set of ordinary differential equations

$$\frac{ds^z}{dt} = -I_1 s s^*$$
$$\frac{ds s^*}{dt} = 2I_1 s^z s s^* \tag{5d.13}$$
$$\frac{dP}{dt} = -2I_1 s^z P.$$

The first two of these equations have a very intuitive physical meaning. They are the trajectories of the classical variables $S^-(t)\, S^+(t)$ and $S^z(t)$ used in subsection (5b) as may be shown by using the ansatz $s s^* = \frac{1}{4}N^2 \sin^2\Phi$ and $s^z = \frac{1}{2}N \cos\Phi$ and comparing with (5b.15). The classical trajectories (5b.17) may be written as

$$s^z = s^z(s_0 s_0^*, s_0^z, t) = \frac{s_0^z - \sqrt{(s_0^z)^2 + s_0 s_0^*}\,\tanh\!\left(I_1 t\sqrt{(s_0^z)^2 + s_0 s_0^*}\right)}{1 - s_0^z[(s_0^z)^2 + s_0 s_0^*]^{-1/2}\tanh\!\left(I_1 t\sqrt{(s_0^z)^2 + s_0 s_0^*}\right)}$$

$$(5d.14)$$

$$s s^* = s s^*(s_0 s_0^*, s_0^z, t) = \frac{s_0 s_0^*\,\mathrm{sech}^2\!\left(I_1 t\sqrt{(s_0^z)^2 + s_0 s_0^*}\right)}{\left[1 - s_0^z[(s_0^z)^2 + s_0 s_0^*]^{-1/2}\tanh\!\left(I_1 t\sqrt{(s_0^z)^2 + s_0 s_0^*}\right)\right]^2}\,.$$

In the following we will also need the inversions of Eqs. (5d.14) expressing the initial coordinates s_0^z and $s_0 s_0^*$ in terms of the current values at time t

$$s_0^z = s_0^z(s s^*, s^z, t) = \frac{s^z + \sqrt{(s^z)^2 + s s^*}\,\tanh\!\left(I_1 t\sqrt{(s^z)^2 + s s^*}\right)}{1 + s^z[(s^z)^2 + s s^*]^{-1/2}\tanh\!\left(I_1 t\sqrt{(s^z)^2 + s s^*}\right)}\,,$$

$$(5d.15)$$

$$s_0 s_0^* = s_0 s_0^*(s s^*, s^z, t) = \frac{s s^*\,\mathrm{sech}^2\!\left(I_1 t\sqrt{(s^z)^2 + s s^*}\right)}{\left[1 + s^z[(s^z)^2 + s s^*]^{-1/2}\tanh\!\left(I_1 t\sqrt{(s^z)^2 + s s^*}\right)\right]^2}\,.$$

From the third of Eqs. (5d.13) we find the quasiprobability

$$P(s s^*, s^z, t) = \frac{s_0 s_0^*(s s^*, s^z, t)}{s s^*}\,P(s_0 s_0^*(s s^*, s^z, t), s_0^z(s s^*, s^z, t), 0)\,. \qquad (5d.16)$$

We thus see that the quasiprobability drifts through the phase space of its independent variables along the classical trajectories. The shape of P does not change in time save for the occurrence of the kinematical factor $s_0 s_0^*/s s^*$. The latter shows how the differential phase space volume element changes along the trajectories. Its presence guarantees that P remains normalized to unity at all times if it was so normalized initially. The precise statement is

$$\mathrm{d}s_0^z\,\mathrm{d}s_0 s_0^* = \frac{s_0 s_0^*(s s^*, s^z, t)}{s s^*}\,\mathrm{d}s^z\,\mathrm{d}s s^*\,. \qquad (5d.17)$$

It may be verified by evaluating the functional determinant for the transformation of variables s_0^z, $s_0 s_0^* \to s^z, s s^*$ given in (5d.15).

Let us now use the result (5d.16) to find the expectation values $\langle S^+(t)^l\, S^z(t)^k\, S^-(t)^{l'}\rangle$. With the help of the general expression (5d.2) we get

$$\langle S^+(t)^l\, S^z(t)^k\, S^-(t)^{l'}\rangle$$

$$= \delta_{ll'}\int_0^\infty \mathrm{d}s s^* \int_{-\infty}^{+\infty} \mathrm{d}s^z (s^z)^k (s s^*)^l\,\frac{s_0 s_0^*(s s^*, s^z, t)}{s s^*}\,P(s_0 s_0^*(s s^*, s^z, t), s_0^z(s s^*, s^z, t), 0)$$

$$= \delta_{ll'}\int_0^\infty \mathrm{d}s_0 s_0^* \int_{-\infty}^{\infty} \mathrm{d}s_0^z\,[s^z(s_0 s_0^*, s_0^z, t)]^k\,[s s^*(s_0 s_0^*, s_0^z, t)]^l\,P(s_0 s_0^*, s_0^z, 0)\,. \qquad (5d.18)$$

In the second line we have transformed the integration variables according to (5d.14) and used (5d.17). The expression obtained allows the following intuitive interpretation. Either the time-independent "random" variables $(s^z)^k (ss^*)^l$ are averaged with the time-dependent quasiprobability distribution as a statistical weight or the time-dependent random variables $(s^z(t))^k (ss^*(t))^l$ are averaged with the initial quasiprobability as statistical weight. The explicit evaluation of (5d.18) requires the specification of the initial quasiprobability distribution.

δ) Initial Quasiprobabilities

For our initial state (5c.10) we have

$$\langle \tfrac{1}{2}N, m|S^{+l} S^{zk} S^{-l'}|\tfrac{1}{2}N, m\rangle = \delta_{ll'} \begin{cases} \dfrac{(\frac{1}{2}N+m)!\,(\frac{1}{2}N-m+l)!}{(\frac{1}{2}N-m)!\,(\frac{1}{2}N+m-l)!}(m-l)^k & \text{for} \quad l \leqq \frac{1}{2}N+m \\ 0 & \text{for} \quad l > \frac{1}{2}N+m\,. \end{cases} \tag{5d.19}$$

By expanding the exponential functions in the definition (5d.1) of the quasiprobability we get the initial quasiprobability as

$$P(ss^*, s^z, 0) = \sum_{l=0}^{\frac{1}{2}N+m} \frac{(-1)^l}{l!} \frac{(\frac{1}{2}N+m)!\,(\frac{1}{2}N-m+l)!}{(\frac{1}{2}N-m)!\,(\frac{1}{2}N+m-l)!} \cdot \delta(s^z+l-m) \frac{\partial^l}{\partial ss^{*l}} \delta(ss^*)\,. \tag{5d.20}$$

The expressions (5d.19) and (5d.20) simplify considerably in the limit $N \gg 1$, since the ratios of factorials may be approximated by

$$(z+a)!/z! \approx z^a \quad \text{for} \quad z \gg a\,. \tag{5d.21}$$

The resulting asymptotic expressions look somewhat different for strong $(\frac{1}{2}N - m \ll \frac{1}{2}N)$, medium $(|m| \ll N/2)$, and weak $(\frac{1}{2}N + m \ll \frac{1}{2}N)$ initial atomic excitation. In the case of strong excitation (5d.20) becomes

$$\begin{aligned} P(ss^*, s^z, 0) &= \delta(s^z - \tfrac{1}{2}N + v) \sum_l (-1)^l (N-v)^l \frac{(v+l)!}{v!} \frac{\partial^l}{\partial ss^{*l}} \delta(ss^*) \\ &= \delta(s^z - \tfrac{1}{2}N + v)(1/v!)[ss^*/(N-v)]^v (N-v)^{-1} \exp[-ss^*/(N-v)] \end{aligned} \tag{5d.22}$$

for $0 \leqq v \equiv \frac{1}{2}N - m \ll \frac{1}{2}N$.

Since $P(ss^*, s^z, 0)$ factors with respect to the dependence on ss^* and s^z we may say that the "random variables" ss^* and s^z are statistically uncorrelated in this strong-excitation initial state. Let us note that v need not be a small number of order unity for the asymptotic expression

(5d.22) to be valid. If N is say 10^{12} and if we are willing to accept an accuracy of 1% for the moments $\langle S^{+l} S^{-l} \rangle$ with $0 \leqq l \leqq 10^8$ then we may use Eq. (5d.22) for v up to $\sim 10^{10}$. For v that large $P(ss^*, s^z, 0)$ develops an extremely sharp maximum at $ss^* = v(N-v) = \frac{1}{4}N^2 - m^2$. Up to corrections of order v/N (that is, in the above example, 1%) for the moments we can then replace (5d.22) with

$$P(ss^*, s^z, 0) = \delta(s^z - m)\,\delta(ss^* - \tfrac{1}{4}N^2 + m^2) \text{ for } 1 \ll v = \tfrac{1}{2}N - m \ll \tfrac{1}{2}N\,. \qquad (5d.23)$$

For medium initial excitation (5d.20) simplifies to

$$P(ss^*, s^z, 0) = \sum_l \frac{(-1)^l}{l!} (\tfrac{1}{4}N^2 - m^2)^l\, \delta(s^z + l - m) \frac{\partial^l}{\partial s s^{*l}} \delta(ss^*) \qquad (5d.24)$$

for $|m| \ll N/2$.

This expression remains valid for $|m|$ up to $\sim 10^{10}$ in the above numerical example. If $|m|$ is large compared to unity this formula simplifies further according to $\delta(s^z + l - m) \to \delta(s^z - m)$ and then coincides with (5d.23). We thus see that the asymptotic expressions for strong and medium initial excitation (5d.22) and (5d.24), respectively, have overlapping ranges of validity. Let us remark that $P(ss^*, s^z, 0)$ according to (5d.24) does not factor in separate distribution functions for its independent variables. These latter are thus "statistically correlated random variables" for medium excitation. It is also interesting that this $P(ss^*, s^z, 0)$ is sharply peaked near $ss^* = \frac{1}{4}N^2 - m^2$ in the following sense

$$\begin{aligned} P(ss^*, t=0) &= \int_{-\infty}^{+\infty} ds^z\, P(ss^*, s^z, 0) \\ &= \sum_l \frac{(-1)^l}{l!} (\tfrac{1}{4}N^2 - m^2)^l \frac{\partial^l}{\partial s s^{*l}} \delta(ss^*) \qquad (5d.25) \\ &= \delta(ss^* - \tfrac{1}{4}N^2 + m^2) \end{aligned}$$

for $|m| \ll N/2$.

A similar formula for P in the case of weak initial excitation is also readily written down but is of no interest for our discussion of superradiance.

ε) Explicit Results

We are now equipped with the initial quasiprobabilities and can therefore explicitly evaluate the integrals (5d.18). We will write down the field expectation values $\langle b^\dagger(t)^l\, b(t)^l \rangle$ which can be measured in photon counting experiments [67]. By using the adiabatic correspondence (5c.22) and Eq. (5d.18) we find the theoretical predictions for $\langle b^\dagger(t)^l\, b(t)^l \rangle$.

In the case of medium initial excitation we use (5d.24) and get

$$\begin{aligned}\langle b^\dagger(t)^l\, b(t)^l\rangle &= |\mu|^{2l} \sum_j \frac{(-1)^j}{j!}\left(\tfrac{1}{4}N^2-m^2\right)^j \\ &\cdot \int_0^\infty \mathrm{d}s_0 s_0^* \left[s s^*(s_0 s_0^*, m-j, t)\right]^l \frac{\partial^j}{\partial s_0 s_0^{*j}}\,\delta(s_0 s_0^*)\,,\end{aligned} \tag{5d.26}$$

where we have to insert the classical trajectory (5d.14). The integral occurring here collects its only important contributions near $s_0 s_0^* \approx N^2/4 - m^2$. For such values of $s_0 s_0^*$ the j-dependence of $s s^*(s_0 s_0^*, m-j, t)$ is negligible to within corrections of order $1/N$. We then get with the help of (5d.25)

$$\begin{aligned}\langle b^+(t)^l\, b(t)^l\rangle &= |\mu|^{2l}\left[s s^*\left(\tfrac{1}{4}N^2-m^2, m, t\right)\right]^l \\ &= \left[|\mu|^2 \tfrac{1}{4}N^2 \operatorname{sech}^2[(t-t_{\max})/\tau]\right. \\ &= \langle b^\dagger(t)\, b(t)\rangle^l\end{aligned} \tag{5d.27}$$

with $t_{\max} = \frac{1}{2}\tau \ln[(\frac{1}{2}N+m)/(\frac{1}{2}N-m)]$ for $|m| \ll N/2$.

This is precisely the result (5b.18) of the semiclassical theory. We get the same result for strong initial excitations, as is clear from (5d.23), as long as $1 \ll \nu = N/2 - m \ll N/2$.

Deviations from the completely classical behavior of superradiant pulses, i.e., noticeable quantum fluctuations can only be expected for very strong initial excitations, i.e. $\nu = \frac{1}{2}N - m = O(1)$. By replacing $N-\nu$ by N in (5d.22) we get for this case

$$\begin{aligned}\langle b^\dagger(t)^l\, b(t)^l\rangle &= |\mu|^{2l} \int_0^\infty \mathrm{d}s_0 s_0^* (1/\nu!)\,(s_0 s_0^*/N)^\nu N^{-1}\, \mathrm{e}^{-s_0 s_0^*/N} \\ &\cdot \left[\frac{s_0 s_0^* \operatorname{sech}\left(I_1 t \sqrt{\frac{1}{4}N^2 + s_0 s_0^*}\right)}{1 - \frac{1}{2}N\left(\frac{1}{4}N^2 + s_0 s_0^*\right)^{-1/2} \tanh\left(I_1 t\sqrt{\frac{1}{4}N^2+s_0 s_0^*}\right)}\right]^l .\end{aligned} \tag{5d.28}$$

Since the important contributions to this integral arise from the interval $0 \leqq s_0 s_0^* \lesssim \nu N$ where $s_0 s_0^* \ll N^2/4$ we may replace $N^2/4 + s_0 s_0^*$ by $N^2/4$ in the arguments of the hyperbolic functions, again accepting an error of order $1/N$. By finally expressing the hyperbolic functions in terms of exponentials we obtain

$$\begin{aligned}\langle b^\dagger(t)^l\, b(t)^l\rangle &= |\mu|^{2l} \int_0^\infty \mathrm{d}s_0 s_0^* (1/\nu!)\,(s_0 s_0^*/N)^\nu N^{-1}\, \mathrm{e}^{-s_0 s_0^*/N} \\ &\cdot \left[\frac{s_0 s_0^* \mathrm{e}^{2t/\tau}}{1 + (s_0 s_0^*/N)\, \mathrm{e}^{2t/\tau}}\right] .\end{aligned} \tag{5d.29}$$

For large ν this again reduces to the classical result (5d.27). We expect the deviations of (5d.29) from the classical result (5d.27) to be most pronounced for $v=0$, since for this case of complete excitation of all atoms the distribution function for the variable $s_0 s_0^*$ is broadest. This special case has also been investigated by Degiorgio [55]. The present expression (5d.29) is valid for v ranging from zero up to values where the system behaves fully classically. In order to evaluate the magnitude of the quantum fluctuations displayed by the superradiant pulse for initial states with $v=0, 1, 2, 3, \ldots$ we first rescale the variables by writing

$$z=N\mathrm{e}^{-2t/\tau}, \qquad s_0 s_0^* = N(y-z)$$

$$\frac{\langle b^\dagger(t)^l\, b(t)^l\rangle}{|\mu|^{2l}(N^2/4)^l} = \frac{(4z)^l \mathrm{e}^z}{v!} \int\limits_z^\infty \mathrm{d}y \frac{(y-z)^{l+v}}{y^{zl}} \mathrm{e}^{-y}. \tag{5d.30}$$

This shows that the number of atoms N enters as a scaling parameter only, once N is large. We compare the results of a numerical evaluation of (5d.30) with the semiclassical results (5b.18) in Fig. 1–3. Figure 1 presents a plot of the "time" z_v at which the pulse intensity goes through its maximum versus the initial-excitation parameter $v=\frac{1}{2}N-m$. The largest deviation from the semiclassical value $z_v^{(\mathrm{cl})}=v/(1-v/N)\approx v$ appears for $v=0$, i.e. full initial excitation of all atoms. For v increasing the relative deviation $(z_v - z_v^{(\mathrm{cl})})/z_v^{(\mathrm{cl})}$ approaches zero as $1/2v$. In Fig. 2 we show the relative deviation of the quantum-mechanically calculated maximum intensity ($l=1, z=z_v$) from the classical maximum intensity $|\mu|^2 N^2/4$

$$\Delta I(v) = 1 - \langle b^\dagger(t)\, b(t)\rangle/(\tfrac{1}{2}N)^2 . \tag{5d.31}$$

The maximum intensity is found smaller than what the classical treatment predicts. The relative deviation is 22% for $v=0$ and approaches $4/(v+1)$

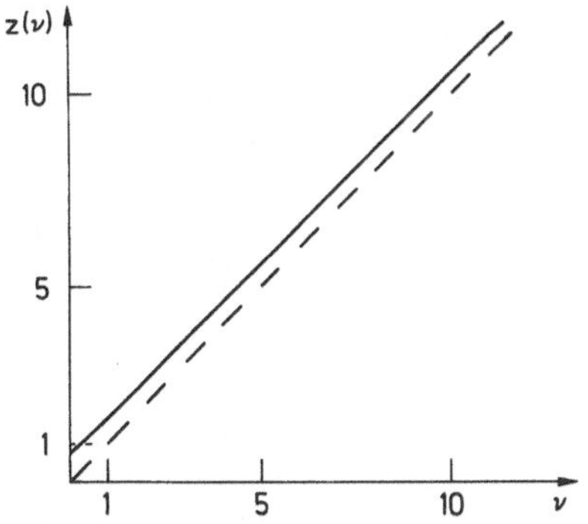

Fig. 1

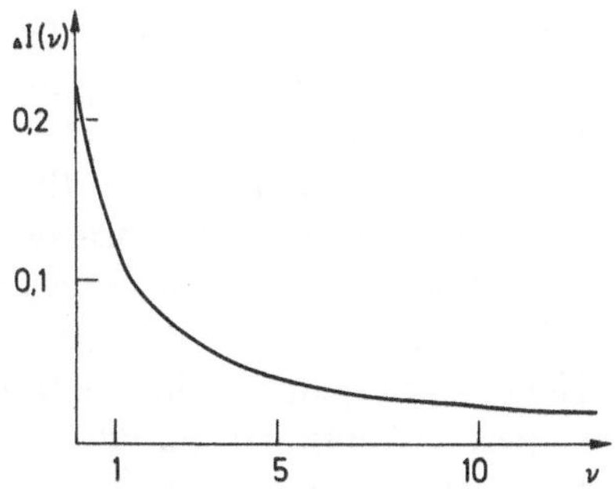

Fig. 2

Fig. 1. Times of maximum intensity, normalized as $z=N\mathrm{e}^{-2t/\tau}$, for atomic initial states $|\frac{1}{2}N, \frac{1}{2}N-v\rangle$. The dashed line gives the classical result $z_v^{(\mathrm{cl})}=v/(1-v/N)\approx v$ according to (5b.19).

Fig. 2. Relative deviation of the maximum intensity from its classical value for atomic initial states $|\frac{1}{2}N, \frac{1}{2}N-v\rangle$. For $v>8$ the curve approaches $1/(v+1)$

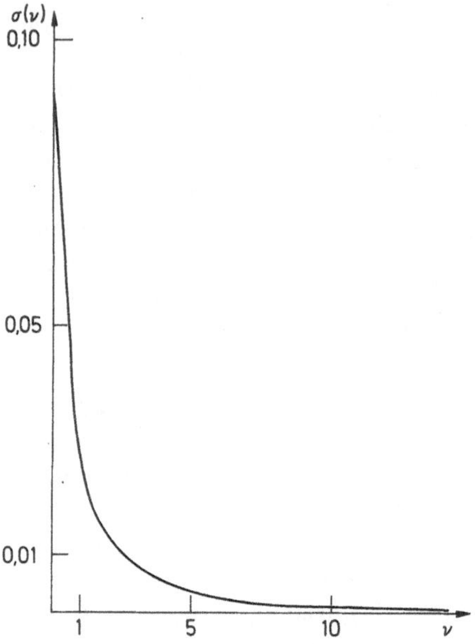

Fig. 3. Quadratic fluctuation of the intensity for atomic initial states $|\frac{1}{2}N, \frac{1}{2}N-\nu\rangle$, evaluated for the times z_ν of maximum intensity. For $\nu > 10$ the curve approaches $1/9\nu^2$

for ν increasing. Finally, in Fig. 3 we give the quadratic fluctuation of the intensity evaluated for the time z_ν of maximum intensity

$$\sigma(\nu) = \left.\frac{\langle b^\dagger(t)\, b^\dagger(t)\, b(t)\, b(t)\rangle - \langle b^\dagger(t)\, b(t)\rangle^2}{\langle b^\dagger(t)\, b(t)\rangle^2}\right|_{z=z_\nu}. \tag{5d.32}$$

For $\nu = 0$ we have $\sigma(0) \approx 0.09$. The pulse displays large quantum fluctuations. These fluctuations rapidly decrease with ν increasing, $\sigma(\nu)$ approaching $1/9\nu^2$.

As a conclusion we may thus say that superradiant pulses behave practically classically for nearly all atomic initial states $|\frac{1}{2}N, m\rangle$ if the number of atoms N is large. An exception is made by the most highly excited initial states with $\nu = \frac{1}{2}N - m = O(1)$ only, for which the pulses display large quantum fluctuations. For $N \to 0$ the domain $0 \leqq \nu \lesssim O(1)$ of these exceptional initial states becomes asymptotically small in relative terms, $O(1)/N \to 0$. The exceptional behavior of the most highly excited atomic initial states is easily understood qualitatively. For these states the pulses are triggered by elementary spontaneous-emission acts which are uncorrelated with each other and may be considered quantum noise. The pulses thus generated can be understood as amplified noise. On the other hand, for initial states with ν large, the atoms initially produce a nonzero transverse component of the Bloch vector, $[\langle S^+(0)\, S^-(0)\rangle]^{1/2} = [(\nu+1)(N-\nu)]^{1/2} \gg 1$. Since the transverse component of the Bloch vector measures the electric polarization of the atomic medium, the pulses are in these cases triggered by a large "classical" source rather than by noise and thus should behave nearly classically as in fact they do.

6. The Laser

6a) Introductory Remarks

As is well known, in a laser stimulated emission of light by excited atoms is used to generate selfsustained oscillations of the electromagnetic field. To achieve this the active atoms have to be pumped continuously to suitable excited states and the radiated field has to be fed back into the atoms by means of mirrors. Laser theory has to account for the atom-field interaction as well as the irreversible pump mechanism and field losses by diffraction and leakage through the non-ideal mirrors. Several formal techniques have been successfully employed to treat the dynamics of a laser. Among these are Langevin equation methods, generalized Fokker Planck equations, master equations, and Green's functions. For an exhaustive presentation of the various equivalent laser theories we refer to [61]. Here we briefly discuss a master equation treatment. For the sake of simplicity we consider the simplest model of a laser consisting of N identical two-level atoms in resonance with a single mode of the electromagnetic field in the cavity. The atom-field density operator $W(t)$ obeys an equation of motion reading, in the interaction picture,

$$\dot{W}(t) = -i(L_{AF} + i\Lambda_F + i\Lambda_A)\, W(t)\,. \tag{6a.1}$$

The three parts of the Liouvillian refer to the atom-field interaction (L_{AF}), the pump and atomic losses (Λ_A) and field losses (Λ_F). They have all been used in the preceding sections of this paper in other contexts.

$$\begin{aligned} L_{AF}X &= \hbar^{-1}[H_{AF}, X]\,, \qquad H_{AF} = \hbar g(bS^+ + b^\dagger S^-) \\ &\qquad\qquad\qquad\qquad\quad = \hbar g \sum_{\nu=1}^{N} (bs_\nu^+ + b^\dagger s_\nu^-) \\ \Lambda_F X &= \kappa\{[b, Xb^\dagger] + [bX, b^\dagger]\} \\ \Lambda_A &= \sum_{\nu=1}^{N} \Lambda_\nu \\ \Lambda_\mu X &= \tfrac{1}{2}\gamma_{10}\{[s_\mu^-, Xs_\mu^+] + [s_\mu^- X, s_\mu^+]\} \\ &\quad + \tfrac{1}{2}\gamma_{01}\{[s_\mu^+, Xs_\mu^-] + [s_\mu^+ X, s_\mu^-]\} \\ &\quad - \tfrac{1}{2}\eta\{[s_\mu^z, Xs_\mu^z] + [s_\mu^z X, s_\mu^z]\}\,. \end{aligned} \tag{6a.2}$$

H_{AF} is the interaction Hamiltonian already used in our treatment of superradiance. Λ_F is the Liouvillian for a damped harmonic oscillator derived in Section 3. It describes how the field mode in the cavity would dissipate any initial energy were it not coupled to the active atoms. Note that we have dropped the term proportional to $\bar{n}=(e^{\beta\hbar\omega}-1)^{-1}$ occurring in (3b.7). This is possible since for frequencies in the optical region and for temperatures at which lasers are usually operated we have $\bar{n}\ll 1$. Finally, Λ_μ is the atomic pump and loss Liouvillian (3c.1). According to (3c.3) the transition rates γ_{10}, γ_{01}, and η are related to the polarization and inversion damping constants $\gamma_\perp$ and $\gamma_\parallel$, respectively and the unsaturated inversion σ_0 as

$$\begin{aligned} T_2^{-1}&=\gamma_\perp=\tfrac{1}{2}(\gamma_{01}+\gamma_{10}+\eta)\\ T_1^{-1}&=\gamma_\parallel=\gamma_{01}+\gamma_{10}\\ \sigma_0&=(\gamma_{01}-\gamma_{10})/(\gamma_{01}+\gamma_{10})\,,\qquad -1\leqq\sigma_0\leqq+1\,. \end{aligned} \tag{6a.3}$$

The Bose commutation relations for the field operators b and $b^\dagger$ and the spin commutation relations for the atomic operators $s_\mu^\pm$ (polarization) and s_μ^z (inversion) are listed in (5e.1).

As discussed in Section 5c the Liouvillians L_{AF} and Λ_F impose on observables of the system the time rates of change $g\sqrt{N}$ and κ, respectively. By the definition of Λ_A in the form

$$\begin{aligned} \mathrm{tr}_A e^{\Lambda_A t}X&=\mathrm{tr}_A X\\ \mathrm{tr}_A s_\mu^\pm e^{\Lambda_A t}X&=e^{-\gamma_\perp t}\mathrm{tr}_A X\\ \mathrm{tr}_A s_\mu^z e^{\Lambda_A t}X&=e^{-\gamma_\parallel t}\mathrm{tr}_A s_\mu^z X+(1-e^{-\gamma_\parallel t})\tfrac{1}{2}\sigma_0\,\mathrm{tr}_A X \end{aligned} \tag{6a.4}$$

we see that Λ_A introduces time rates of change of the order $\gamma_\perp$ and $\gamma_\parallel$. We thus have the important order-of-magnitude estimates

$$\begin{aligned} O(\Lambda_A)&=\gamma_\perp,\gamma_\parallel\\ O(\Lambda_F)&=\kappa\\ O(L_{AF})&=g\sqrt{N}\,. \end{aligned} \tag{6a.5}$$

Let us note that in our discussion of superradiance in Section 5c we have solved Eq. (6a.1) in the limit $O(\Lambda_A)\ll O(L_{AF})\ll O(\Lambda_F)$. Here, however, the light field is trapped in a near-ideal cavity so that we have $O(\Lambda_F)\ll O(\Lambda_A), O(L_{AF})$. Moreover, for a typical gas laser Arecchi et al. [68] give $\gamma_\perp\approx\gamma_\parallel$ and $g\sqrt{N}/\gamma_\perp\approx 0.1$. We thus have to solve Eq. (6a.1) in the limit

$$\kappa\ll g\sqrt{N}<\gamma_\perp,\gamma_\parallel\,. \tag{6a.6}$$

This suggests to first try to eliminate the atomic variables from (6a.1) and to consider a Nakajima-Zwanzig equation for the reduced density operator $\varrho(t)$ of the field mode. Because $O(L_{AF})/O(\Lambda_A) \approx 0.1$ is smaller than unity but not very small we cannot, in general, hope a low-order approximation of the Nakajima-Zwanzig equation in terms of L_{AF} to be possible. It will turn out, however, that for the laser operating near threshold such a low-order approximation is indeed valid. We will treat this simple case only.

6b) Master Equation for the Field Density Operator

In order to eliminate the atomic variables from (6a.1) we use the projector

$$\mathfrak{P} = A\,\mathrm{tr}_A \tag{6b.1}$$

and use as the reference state $B_{\mathrm{ref}} \equiv A$ for the atoms the unsaturated atomic density operator

$$\Lambda_A A = 0\,, \quad A = \prod_{\mu=1}^{N} A_\mu\,, \quad A_\mu = \tfrac{1}{2}(1-\sigma_0)\, s_\mu^- s_\mu^+ + \tfrac{1}{2}(1+\sigma_0)\, s_\mu^+ s_\mu^-\,. \tag{6b.2}$$

We will have to demonstrate later that this choice for the atomic reference state is a good one for a laser operating near threshold. We now consider the Nakajima-Zwanzig equation (2b.13) for

$$\varrho(t) = \mathrm{tr}_A\, W(t)\,, \tag{6b.3}$$

$$\dot{\varrho}(t) = \Lambda_F \varrho(t) + \int_0^t \mathrm{d}t'\, K(t')\, \varrho(t-t') + I(t)\,. \tag{6b.4}$$

Because of

$$\mathrm{tr}_A (L_{AF})^{2n+1} A = 0\,, \qquad n = 0, 1, 2, \ldots \tag{6b.5}$$

and since $O(\Lambda_F) \ll O(\Lambda_A)$ the expansion (2b.18) for the Laplace-transformed integral kernel reads here

$$K(z) = \sum_{n=0}^{\infty} K^{(2n+2)}(z) = \sum_{n=0}^{\infty} (-1)^{n+1}\, \mathrm{tr}_A\, L_{AF} [U(z)\,(1-\mathfrak{P})\, L_{AF}]^{2n+1} A$$

with

$$U(z) = \int_0^{\infty} \mathrm{d}t\, \mathrm{e}^{-zt}\, \mathrm{e}^{\Lambda_A t} = (z - \Lambda_A)^{-1}\,. \tag{6b.6}$$

A similar expansion obtains for the inhomogeneity $I(t)$. For the further evaluation of (6b.6) it is convenient to introduce the diagonal representation of $\varrho(t)$ with respect to coherent states

$$\varrho(t) = \int \mathrm{d}^2\beta\, P(\beta, \beta^*, t)\, |\beta\rangle \langle\beta|\,, \quad b|\beta\rangle = \beta|\beta\rangle \tag{6b.7}$$

which we have already used in Sections 3 and 5. (6b.4) then becomes an integrodifferential equation of motion for the quasiprobability $P(\beta, \beta^*, t)$. This equation has the same appearance as (6b.4) with $\varrho \to P$ and [7]

$$\begin{aligned} \Lambda_F &= \kappa \left(\frac{\partial}{\partial \beta^*} \beta^* + \frac{\partial}{\partial \beta} \beta \right) \\ L_{AF} X &= g \left\{ \beta [S^+, X] + \beta^* [S^-, X] + \frac{\partial}{\partial \beta^*} X S^+ - \frac{\partial}{\partial \beta} S^- X \right\}. \end{aligned} \tag{6b.8}$$

Then the first two terms in the expansion (6b.6) read [70]

$$K^{(2)}(t) = \varphi^{(2)}(t) \, (N g^2/\gamma_\perp) \left\{ -\left(\frac{\partial}{\partial \beta^*} \beta^* + \frac{\partial}{\partial \beta} \beta \right) \sigma_0 + \frac{\partial^2}{\partial \beta^* \partial \beta} (1 + \sigma_0) \right\}$$

and (6b.9)

$$\begin{aligned} K^{(4)}(t) = {} & (4 N g^4 \sigma_0 / \gamma_\perp^2 \gamma_\parallel) \left(\frac{\partial}{\partial \beta^*} \beta^* + \frac{\partial}{\partial \beta} \beta \right) \beta^* \beta \varphi_1^{(4)}(t) \\ & + [4 N g^4 (1 + \sigma_0) / \gamma_\perp^2 \gamma_\parallel] \left(\frac{\partial}{\partial \beta^*} \beta^* + \frac{\partial}{\partial \beta} \beta \right) \varphi_1^{(4)}(t) \\ & + (N^2 g^4 / \gamma_\perp^3) \left[\frac{\partial^2}{\partial \beta^* \partial \beta} \{ \tfrac{1}{2} \sigma_0 (1 + \sigma_0) \varphi_2^{(4)}(t) \right. \\ & - (2 \gamma_\perp / \gamma_\parallel N) (1 + \sigma_0)(2 + \sigma_0) \varphi_1^{(4)}(t) \} \\ & + \left(\frac{\partial^2}{\partial \beta^{*2}} \beta^{*2} + 2 \frac{\partial^2}{\partial \beta^* \partial \beta} \beta^* \beta + \frac{\partial^2}{\partial \beta^2} \beta^2 \right) \\ & \cdot \{ \tfrac{1}{2} \sigma_0^2 \varphi_2^{(4)}(t) - (2 \gamma_\perp / \gamma_\parallel N) \varphi_1^{(4)}(t) \} \\ & - \frac{\partial^2}{\partial \beta^* \partial \beta} \left(\frac{\partial}{\partial \beta^*} \beta^* + \frac{\partial}{\partial \beta} \beta \right) (1 + \sigma_0) \\ & \cdot \{ \sigma_0 \varphi_2^{(4)}(t) - (\gamma_\perp / \gamma_\parallel N) \varphi_1^{(4)}(t) \} \\ & \left. + \frac{\partial^4}{\partial \beta^{*2} \partial \beta^2} (1 + \sigma_0)^2 \{ \tfrac{1}{2} \varphi_2^{(4)}(t) + (2 \gamma_\perp / \gamma_\parallel N) \varphi_1^{(4)}(t) \} \right]. \end{aligned} \tag{6b.10}$$

Here we have approximated $(N \pm 1)$ by N. The time-dependence of $K^{(2)}(t)$ and $K^{(4)}(t)$ is determined by the "retardation functions"

$$\begin{aligned} \varphi^{(2)}(t) &= \gamma_\perp e^{-\gamma_\perp t} \\ \varphi_1^{(4)}(t) &= [\gamma_\perp^2 \gamma_\parallel / (\gamma_\perp - \gamma_\parallel)^2] \, [e^{-\gamma_\parallel t} - \{1 + (\gamma_\perp - \gamma_\parallel) t\} e^{-\gamma_\perp t}] \\ \varphi_2^{(4)}(t) &= 2 \gamma_\perp [e^{-2\gamma_\perp t} - \{1 - \gamma_\perp t\} e^{-\gamma_\perp t}]. \end{aligned} \tag{6b.11}$$

[7] This can be shown by writing Eq. (6b.4) in antinormal order with respect to b and $b^\dagger$ using $[b, f(b, b^\dagger)] = \partial f(b, b^\dagger)/\partial b^\dagger$ and then substituting $b \to \beta$, $b^\dagger \to \beta^*$, $\varrho \to P$ [69].

These functions are normalized so as to integrate up to unity, $\int_0^\infty dt\,\varphi(t)=1$. We will come back to the higher order terms $K^{(2n+2)}(t)$ below. The first order derivative terms in $K^{(2)}(t)$ describe a linear drift of the quasiprobability $P(\beta, \beta^*, t)$ towards higher amplitudes $|\beta|$, i.e. a linear gain for the field amplitude. The amplification coefficient is

$$\alpha_l = Ng^2\sigma_0/\gamma_\perp . \tag{6b.12}$$

This amplification competes with the linear damping described by Λ_F. The laser can begin to produce selfsustained oscillations once the linear gain outweighs the linear damping,

$$\alpha_l > \kappa \Rightarrow \sigma_0 > \sigma_{\text{thr}} = \frac{\kappa\gamma_2}{Ng^2} . \tag{6b.13}$$

This is the wellknown threshold condition [61]. The second-order-derivative term in $K^{(2)}(t)$ has a diffuse effect on the quasiprobability $P(\beta, \beta^*, t)$. That means physically, it describes noise. The diffusion constant is

$$4q = 4Ng^2(1+\sigma_0)/\gamma_\perp . \tag{6b.14}$$

The first term in $K^{(4)}(t)$ represents a nonlinear damping force on the field amplitude. This is a saturation effect preventing the field amplitude from blowing up for $\alpha_l > \kappa$ and ensuring stable selfsustained oscillations above treshold. The nonlinear-drift coefficient is

$$\alpha_{nl} = \frac{4Ng^4\sigma_0}{\gamma_\perp^2\gamma_\parallel} . \tag{6b.15}$$

Let us now determine, by a selfconsistent argument, the equation of motion for $P(\beta, \beta^*, t)$ near threshold, i.e. for the case

$$\alpha_l \gtrsim \kappa \Rightarrow \sigma_0 \gtrsim \sigma_{\text{thr}} = \kappa\gamma_\perp/Ng^2 . \tag{6b.16}$$

To this end we first assume, subject to later proof, that we can neglect all $K^{(2n+2)}(t)$ for $n>1$, moreover all terms in $K^{(4)}(t)$ except the first, the inhomogeneity $I(t)$ and retardation effects. Then Eq. (6b.4) simplifies to

$$\begin{aligned} &\dot{P}(\beta, \beta^*, t) = \Lambda P(\beta, \beta^*, t) \\ &\Lambda = \left(\frac{\partial}{\partial\beta^*}\beta^* + \frac{\partial}{\partial\beta}\beta\right)(\kappa - \alpha_l + \alpha_{nl}\beta^*\beta) + 4q\frac{\partial^2}{\partial\beta^*\partial\beta} . \end{aligned} \tag{6b.17}$$

This is a Fokker Planck equation first found by means of semiclassical arguments and solved by Risken [70–72]. In as much as it yields a valid description of the laser it proves the field mode to behave like a noise-driven van der Pol oscillator [73]. Its stationary solution is easily verified to be

$$\bar{P}(\beta, \beta^*) = Z^{-1} \exp[-\tfrac{1}{4}(I-a)^2] \tag{6b.18}$$

with

$$I = \sqrt{\alpha_{nl}/\kappa}\, \beta^* \beta$$

$$a = (\alpha_l - \kappa)(\alpha_{nl} q) \approx \sqrt{N\gamma_{\|}/\kappa}\,(\sigma_0 - \sigma_{\text{thr}})$$

$$1 = \int d^2\beta\, \bar{P}(\beta, \beta^*) .$$

This stationary distribution function and its moments

$$\langle b^{\dagger l} b^l \rangle = \int d^2\beta (\beta^* \beta)^l\, \bar{P}(\beta, \beta^*) \tag{6b.19}$$

give an excellent quantitative account of photon counting experiments [74]. The time-dependent solution of the Fokker Planck equation has been given by Risken and Vollmer [71] and Hampstead and Lax [72]. It allows the evaluation of multitime correlation functions of the field operators b and $b^\dagger$ with the help of our general expression (2e.17). Such correlation functions have been measured in both interference-type and photon-counting experiments and again, excellent agreement between theory and experiments is found [75, 76].

While it is gratifying that the simple Fokker Planck equation (6b.17) checks so well with experiments, the field mode thus behaving like a noise-driven van der Pol oscillator, this fact cannot be considered, from a theoretical point of view, a justification for the above-mentioned approximations leading from the general Nakajima-Zwanzig equation to (6b.17). The justification can, however, be given as follows.

The stationary photon number at threshold ($\sigma_0 = \sigma_{\text{thr}} \Rightarrow a = 0$) follows from (6b.18) and (6b.19) as

$$\langle b^\dagger b \rangle|_{a=0} = \sqrt{q/\alpha_{nl}}\,(2/\sqrt{\pi}) \approx \sqrt{\gamma_{\|} N/\kappa} \gg 1 . \tag{6b.20}$$

The experimental result [74] is $\langle b^\dagger b \rangle|_{a=0} \approx 10^4$. For a laser operating near threshold, $(\sigma_0 - \sigma_{\text{thr}})/\sigma_{\text{thr}} \ll 1$, and for small deviations from the stationary regime $[\langle b^\dagger b \rangle|_{a=0}]^{1/2}$ is a good scale for the field variables β, β^*. By introducing the normalized variables

$$\tilde{\beta} = \beta \sqrt[4]{\alpha_{nl}/q}, \qquad \tilde{\beta}^* = \beta^* \sqrt[4]{\alpha_{nl}/q}, \tag{6b.21}$$

we see that all terms in the Fokker Planck differential operator Λ (6b.17) have the same weight, whereas all neglected terms in $K^{(4)}(t)$ are smaller to at least first order in the parameters $\sqrt[4]{\alpha_{nl}/q}$ or $\sqrt[4]{\alpha_{nl}/q}\, g\sqrt{N}/\gamma_\perp$. The higher order contributions $K^{(2n+2)}(t)$ to the integral kernel turn out to be small in the same sense.

We now have to demonstrate the validity of the Markov approximation made above. That is we have to show that $P(\beta, \beta^*, t)$ relaxes to the stationary state (6b.18) much more slowly than the retardation functions (6b.11) decay to zero. The rate of relaxation of the quasiprobability is given by the eigenvalues of the Fokker Planck differential operator Λ. These have been determined by Risken and Vollmer [71] by solving the eigenvalue problem

$$\begin{aligned} &\Lambda P_{nm} = -\gamma_{nm} P_{nm} \\ &\gamma_{nm} \geqq 0 \qquad n = 0, \pm 1, \pm 2, \ldots \\ &\qquad\qquad m = 0, 1, 2, \ldots . \end{aligned} \tag{6b.22}$$

For all retardation effects to be negligible we have to require

$$\gamma_{nm} \ll \gamma_\perp, \gamma_\parallel . \tag{6b.23}$$

By inspection of the results of [71] we find that this condition is fulfilled near threshold, i.e. for $(\sigma_0 - \sigma_{\text{thr}})/\sigma_{\text{thr}} \ll 1$. Since the inhomogeneity $I(t)$ decays on the same time scale as the integral kernel we now also see that we can indeed neglect it.

As a final check on the consistency of our arguments we should show that the choice (6b.2) for the atomic reference state is a good one. For this to be so A should be practically identical with the stationary atomic density operator $\bar{\varrho}_A = \varrho_A(t \to \infty)$. By using (2b.12) $\bar{\varrho}_A$ can be evaluated in the same approximation ($O(g^4)$ and Markov) as the field density operator. It is thus easily shown that $\bar{\varrho}_A$ has indeed the same structure as A, namely

$$\bar{\varrho}_A = \prod_{\mu=1}^{N} \bar{\varrho}_\mu, \quad \bar{\varrho}_\mu = \tfrac{1}{2}(1 - \bar{\sigma})\, s_\mu^- s_\mu^+ + \tfrac{1}{2}(1 + \bar{\sigma})\, s_\mu^+ s_\mu^- \tag{6b.24}$$

and that the deviation $(\sigma_0 - \bar{\sigma})/\sigma_0$ is small near threshold, i.e. for $(\sigma_0 - \sigma_{\text{thr}})/\sigma_{\text{thr}} \ll 1$.

Let us conclude this section with a few qualitative remarks on the laser operated far away from threshold. To treat this case in the framework of the Nakajima-Zwanzig theory would require to retain terms of all orders in the expansion (6b.6) of the integral kernel. Far above threshold ($\sigma_0 \gg \sigma_{\text{thr}}$) the quasiprobability obeys a generalized Fokker

Planck equation [77]

$$\dot{P}(\beta,\beta^*,t)=\kappa\left(\frac{\partial}{\partial\beta^*}\beta^*+\frac{\partial}{\partial\beta}\beta\right)$$
$$+\int_0^t dt'\left[\left(\frac{\partial}{\partial\beta^*}\beta^*+\frac{\partial}{\partial\beta}\beta\right)D(\beta,\beta^*,t')+\frac{\partial^2}{\partial\beta^{*2}}Q^{(++)}(\beta,\beta^*,t')\right. \tag{6b.25}$$
$$\left.+\frac{\partial^2}{\partial\beta^*\partial\beta}Q^{(+-)}(\beta,\beta^*,t')+\frac{\partial^2}{\partial\beta^2}Q^{(--)}(\beta,\beta^*,t')\right]P(\beta,\beta^*,t-t')$$
$$+I(t),$$

since derivatives of higher order than the second assume an ever smaller weight with the pump strength σ_0 and thus the photon number $\langle b^\dagger b\rangle$ increasing. The drift and diffusion coefficients contain contributions of all orders in the coupling constant g. Far below threshold when only a few photons are present all saturation effects are negligible but derivatives of all orders with respect to the field variables have to be kept. We refrain from treating these cases here quantitatively, since they are more easily handled by other methods [78].

7. Dynamics of Critical Fluctuations in the Heisenberg Magnet

7a) Introductory Remarks

It is known from experiments that the dynamical behavior of systems near critical points is characterized by extremely large scales for both the magnitude and the lifetimes of the fluctuations of certain observables. Among these socalled critical observables are always the long-wavelength Fourier components of the order parameter.

For the Heisenberg magnet we have as a complete set of microscopic observables the wave-vector-dependant spin operators $\hat{S}^\alpha_{\boldsymbol{q}}$ ($\alpha=z,+,-$) which obey the commutation relations

$$\begin{aligned}[][\hat{S}^z_{\boldsymbol{q}},\hat{S}^\pm_{\boldsymbol{q}'}]&=\pm N^{-1/2}\hat{S}^\pm_{\boldsymbol{q}+\boldsymbol{q}'}\\ [\hat{S}^+_{\boldsymbol{q}},\hat{S}^-_{\boldsymbol{q}'}]&=2N^{-1/2}\hat{S}^z_{\boldsymbol{q}+\boldsymbol{q}'}\end{aligned} \tag{7a.1}$$

with N = number of spins in the lattice.

The dynamics of these observables is governed by the Heisenberg Hamiltonian

$$H=-\hbar\sum_{\boldsymbol{q}}J(\boldsymbol{q})\,(\hat{S}^z_{\boldsymbol{q}}\hat{S}^z_{-\boldsymbol{q}}+\hat{S}^+_{\boldsymbol{q}}\hat{S}^-_{-\boldsymbol{q}})\,. \tag{7a.2}$$

$J(\boldsymbol{q})$ is the exchange integral. For it we assume

$$J(x_i - x_i) = \sum_{\boldsymbol{q}} J(\boldsymbol{q}) = 0$$
$$J(\boldsymbol{q}) = J(-\boldsymbol{q})\,. \tag{7a.3}$$

The critical fluctuations of these spin variables have been investigated recently by Resibois and de Leener [79], Resibois and Dewel [80], and Kawasaki [81] with the following results. The decay time of the equilibrium correlation function $\Gamma_{\boldsymbol{q}}(t) = \langle \hat{S}^z_{\boldsymbol{q}}(t)\,\hat{S}^z_{-\boldsymbol{q}}(0)\rangle$ diverges at the Curie point ($T = T_c$) for $\boldsymbol{q} \to 0$ as $|q|^{-5/2}$ to within a possible correction $\eta(\eta \ll 1)$ to the exponent. The decay of $\Gamma_{\boldsymbol{q}}(t)$ is non-Markovian, that is $\Gamma_{\boldsymbol{q}}(t)$ displays damped oscillations. These results imply that at $T = T_c$ the conventional theory of critical slowing down [82, 83] is not valid. This latter theory would predict a spin diffusion according to $\dot{\Gamma}_{\boldsymbol{q}}(t) = -\boldsymbol{q}^2 D \Gamma_{\boldsymbol{q}}(t)$. For $0 < |(T - T_c)/T_c| \ll 1$, however, there is a spin diffusion regime for wavevectors smaller than the inverse correlation length, $|\boldsymbol{q}| \ll 1/\xi(T)$. There the decay of $\Gamma_{\boldsymbol{q}}(t)$ is monotonic on a scale $\approx |\boldsymbol{q}|^{-2}$.

These results were obtained by Resibois et al. by an appropriately renormalized perturbation expansion of $\Gamma_{\boldsymbol{q}}(t)$. Kawasaki, on the other hand, proposed a more widely applicable theory. He put forward general kinetic equations which are nonlinear stochastic equations of motion (Langevin equations) for critical dynamical variables. These kinetic equations generalize the conventional linear damping equations by including couplings between the Fourier components of the critical variables. The validity of Kawasaki's approach is supported by the following facts. (i) The kinetic equations imply the correctness of the "dynamical scaling laws" [84] if the static equilibrium correlations of a system in question obey the "static scaling laws" [85, 86]. There is a wealth of experimental evidence for these scaling laws which the Kawasaki theory can thus claim as a back-up for itself, too. (ii) By accounting for couplings between the critical variables Kawasaki's equations incorporate the mode-mode-coupling theory of Kadanoff and Swift [87] which has proved successful in explaining critical fluctuations in liquid-gas systems. (iii) Similar nonlinear Langevin equations have been fruitfully employed in statistical treatments of turbulence [88, 89]. (iv) For the case of the isotropic Heisenberg magnet the solutions of Kawasaki's equations reproduce the results of Resibois et al.

From a theoretical point of view Kawasaki's theory appears to be a phenomenological one. In constructing it Kawasaki made a number of assumptions which are unproven although partly plausible and backed up by empirical evidence. Among these assumptions are the following. (i) The critical dynamical variables move slowly compared to all other variables of the system. (ii) The quantum mechanical operators representing the critical variables can be treated as c-numbers. (iii) Only quadratic nonlinearities occur in the kinetic equations. (iv) Certain higher order static correlation functions of the critical variables factor into products of low-order correlation functions.

We here want to show that Kawasaki's kinetic equations can be derived from the Liouville-van Neumann equation $\dot{\hat{W}}(t) = -(i/\hbar)[H, \hat{W}(t)]$ without recourse to the a-priori assumptions just mentioned. We will do that for the Heisenberg magnet. Other systems can be treated analogously. Our procedure [90] will be based on associating c-number variables with the spin operators $\hat{S}^\alpha_{\boldsymbol{q}}$ in the sense of Section 2d and writing the Liouville-van Neumann equation as a differential equation of motion for a suitably defined quasiprobability distribution function. We then separate the set of wave-vector-dependant spin variables in long-wavelength and short-wavelength variables and show that only the former undergo critical slowing down. The Nakajima-Zwanzig equation for the reduced quasiprobability distribution over the low $-|\boldsymbol{q}|$ variables is found to be a Fokker Planck equation stochastically equivalent to Kawasaki's Langevin equations.

7b) Master Equation for the Critical Dynamical Variables

α) Quasiprobability Distribution Function

We first define a quasiprobability distribution over all spin variables

$$W(\xi, \xi^*, \eta, t) = \tilde{\mathfrak{F}}\left[\exp - i \sum_{\boldsymbol{q}} \{\xi(\boldsymbol{q})\tilde{\xi}(\boldsymbol{q}) + \xi^*(\boldsymbol{q})\tilde{\xi}^*(\boldsymbol{q}) + \eta(\boldsymbol{q})\tilde{\eta}(\boldsymbol{q})\}\right] \cdot F(\tilde{\xi}, \tilde{\xi}^*, \tilde{\eta}, t) \tag{7b.1}$$

as the Fourier transform of the characteristic function

$$F(\tilde{\xi}, \tilde{\xi}^*, \tilde{\eta}, t) = \operatorname{tr} \hat{E}(\tilde{\xi})\, \hat{E}(\tilde{\eta})\, \hat{E}(\tilde{\xi}^*)\, \hat{W}(t)$$

with

$$\begin{aligned} \hat{E}(\tilde{\xi}) &= \exp i \sum_{\boldsymbol{q}} \tilde{\xi}(\boldsymbol{q})\, \hat{S}^-_q \\ \hat{E}(\tilde{\eta}) &= \exp i \sum_{\boldsymbol{q}} \tilde{\eta}(\boldsymbol{q})\, \hat{S}^z_q \\ \hat{E}(\tilde{\xi}^*) &= \exp i \sum_{\boldsymbol{q}} \tilde{\xi}^*(\boldsymbol{q})\, \hat{S}^+_{-q}\,. \end{aligned} \tag{7b.2}$$

For each value of the wavevector q $\xi(q)$ and $\xi^*(q)$ (and likewise $\tilde{\xi}(q)$ and $\tilde{\xi}^*(q)$) are a pair of complex conjugate variables, whereas we choose $\eta(q) = \eta^*(-q)$ (likewise $\tilde{\eta}(q) = \tilde{\eta}^*(-q)$). The multidimensional integration

$\tilde{\mathfrak{J}}$ goes over the real and imaginary parts of the variables $\tilde{\xi}, \tilde{\xi}^*, \tilde{\eta}$ from $-\infty$ to $+\infty$

$$\tilde{\mathfrak{J}} = \int \left(\prod_{\boldsymbol{q}} \mathrm{d}^2\tilde{\xi}(\boldsymbol{q})/\pi\right) \left[(\mathrm{d}\tilde{\eta}(0)/2\pi) \prod_{\boldsymbol{q} \neq 0} \mathrm{d}^2\tilde{\eta}(\boldsymbol{q})/\pi\right]. \tag{7b.3}$$

The wavevector sum and products cover the region

$$0 \leqq |\boldsymbol{q}| \lesssim 1/a\,, \quad a = \text{lattice constant}\,. \tag{7b.4}$$

The quasiprobability distribution function thus defined is real because of $(\hat{S}^{\pm}_{\boldsymbol{q}})^\dagger = \hat{S}^{\mp}_{-\boldsymbol{q}}$ and $(\hat{S}^z_{\boldsymbol{q}})^\dagger = \hat{S}^z_{-\boldsymbol{q}}$ and has the moments

$$\begin{aligned}&\langle \hat{S}^-_{\boldsymbol{q}_1} \hat{S}^-_{\boldsymbol{q}_2} \dots \hat{S}^-_{\boldsymbol{q}_n} \hat{S}^z_{\boldsymbol{q}'_1} \hat{S}^z_{\boldsymbol{q}'_2} \dots \hat{S}^z_{\boldsymbol{q}'_{n'}} \hat{S}^+_{\boldsymbol{q}''_1} \hat{S}^+_{\boldsymbol{q}''_2} \dots \hat{S}^+_{\boldsymbol{q}''_{n''}} (t) \rangle \\ &= \mathfrak{J} \xi(\boldsymbol{q}_1) \dots \xi(\boldsymbol{q}_n)\, \eta(\boldsymbol{q}'_1) \dots \eta(\boldsymbol{q}'_{n'})\, \xi^*(-\boldsymbol{q}''_1) \dots \xi^*(-\boldsymbol{q}''_{n''})\, W(\xi, \xi^*, \eta, t)\end{aligned} \tag{7b.5}$$

with

$$\mathfrak{J} = \int \left(\prod_{\boldsymbol{q}} \mathrm{d}^2\xi(\boldsymbol{q})\right) \left(\mathrm{d}\eta(0) \prod_{\boldsymbol{q} \neq 0} \mathrm{d}^2\eta(\boldsymbol{q})\right).$$

Let us note that $W(\xi, \xi^*, \eta, t)$ is the many-spin analog of the single-spin distribution function (5d.1) we have used in our discussion of superradiance.

β) Equation of Motion for $W(\xi, \xi^*, \eta, t)$

In order to construct the equation of motion for the quasiprobability distribution we need the following identities which generalize (5d.6) and (5d.7)

$$\begin{aligned}
\hat{S}^+_{\boldsymbol{q}} \hat{E}(\tilde{\xi}^*) &= \frac{\partial}{\partial i\tilde{\xi}^*(-\boldsymbol{q})} \hat{E}(\tilde{\xi}^*) \\
\hat{S}^z_{\boldsymbol{q}} \hat{E}(\tilde{\eta}) &= \frac{\partial}{\partial i\tilde{\eta}(\boldsymbol{q})} \hat{E}(\tilde{\eta}) \\
\hat{S}^+_{\boldsymbol{q}} \hat{E}(\tilde{\xi}) &= \hat{E}(\tilde{\xi}) \hat{S}^+_{\boldsymbol{q}} + 2N^{-1/2} \sum_{\boldsymbol{q}'} i\tilde{\xi}(\boldsymbol{q}') \hat{E}(\tilde{\xi}) \hat{S}^z_{\boldsymbol{q}+\boldsymbol{q}'} \\
&\quad - N^{-1} \sum_{\boldsymbol{q}'\boldsymbol{q}''} i\tilde{\xi}(\boldsymbol{q}') i\tilde{\xi}(\boldsymbol{q}'') \frac{\partial}{\partial i\tilde{\xi}(\boldsymbol{q}+\boldsymbol{q}'+\boldsymbol{q}'')} \hat{E}(\tilde{\xi}) \\
\hat{S}^z_{\boldsymbol{q}} \hat{E}(\tilde{\xi}) &= \hat{E}(\tilde{\xi}) \hat{S}^z_{\boldsymbol{q}} - N^{-1/2} \sum_{\boldsymbol{q}'} i\tilde{\xi}(\boldsymbol{q}') \frac{\partial}{\partial i\tilde{\xi}(\boldsymbol{q}+\boldsymbol{q}')} \hat{E}(\tilde{\xi}) \\
\hat{S}^+_{\boldsymbol{q}} \hat{E}(\tilde{\eta}) &= \hat{E}(\tilde{\eta}) \sum_{\boldsymbol{q}'} (\mathrm{e}^{\alpha})_{\boldsymbol{q},\boldsymbol{q}'} \hat{S}^+_{\boldsymbol{q}'}\,,
\end{aligned} \tag{7b.6}$$

with

$$(\alpha)_{\boldsymbol{q}\boldsymbol{q}'} = -N^{-1/2} i\tilde{\eta}(\boldsymbol{q}' - \boldsymbol{q})\,.$$

By inserting these identities into

$$\dot{F}(\tilde{\xi}, \tilde{\xi}^*, \tilde{\eta}, t) = -(i/\hbar)\,\mathrm{tr}\,\hat{E}(\tilde{\xi})\,\hat{E}(\tilde{\eta})\,\hat{E}(\tilde{\xi}^*)\,[\hat{H}, \hat{W}(t)] \tag{7b.7}$$

we first get a differential equation for the characteristic function F and then, by Fourier transforming according to (7b.1), the desired equation of motion for the quasiprobability distribution function,

$$\dot{W}(\xi, \xi^*, \eta, t) = -iL\,W(\xi, \xi^*, \eta, t), \tag{7b.8}$$

with the "Liouvillian"

$$\begin{aligned} L &= -L^* \\ &= \sum_{\boldsymbol{q}\boldsymbol{q}'} J(\boldsymbol{q}', \boldsymbol{q}'-\boldsymbol{q})\, M(\boldsymbol{q})\, \xi(\boldsymbol{q}')\, \xi^*(\boldsymbol{q}'-\boldsymbol{q}) \\ &\quad + 2N^{-1/2} \sum_{\boldsymbol{q}\boldsymbol{q}'} J(\boldsymbol{q}', \boldsymbol{q}'-\boldsymbol{q}) \left\{ \frac{\partial}{\partial \xi^*(\boldsymbol{q})}\, \xi^*(\boldsymbol{q}')\, \eta(\boldsymbol{q}'-\boldsymbol{q}) - \text{c.c.} \right\} \\ &\quad + N^{-1} \sum_{\boldsymbol{q}\boldsymbol{q}'\boldsymbol{q}''} J(\boldsymbol{q}', \boldsymbol{q}'-\boldsymbol{q}) \\ &\qquad \cdot \left\{ \frac{\partial^2}{\partial \xi^*(\boldsymbol{q})\, \partial \xi^*(\boldsymbol{q}'')}\, \xi^*(\boldsymbol{q}')\, \xi^*(\boldsymbol{q}+\boldsymbol{q}''-\boldsymbol{q}') - \text{c.c.} \right\}, \end{aligned} \tag{7b.9}$$

$$\begin{aligned} M(\boldsymbol{q}) &= \sum_{n=0}^{\infty} \frac{1}{(n+1)!} (1/\sqrt{N})^{n+1} \sum_{\boldsymbol{q}_1 \boldsymbol{q}_2 \ldots \boldsymbol{q}_n} \\ &\quad \cdot \frac{\partial^{n+1}}{\partial \eta(\boldsymbol{q}_1)\, \partial \eta(\boldsymbol{q}_2) \ldots \partial \eta(\boldsymbol{q}_n)\, \partial \eta(\boldsymbol{q}-\boldsymbol{q}_1-\boldsymbol{q}_2 \cdots - \boldsymbol{q}_n)}, \\ J(\boldsymbol{q}', \boldsymbol{q}) &= J(\boldsymbol{q}') - J(\boldsymbol{q}). \end{aligned}$$

γ) Low-$|\boldsymbol{q}|$ and High-$|\boldsymbol{q}|$ Variables

Let us now separate the set of spin variables $\{\xi, \xi^*, \eta\}$ in long-wavelength and short-wavelength variables as follows

$$\xi, \xi^*, \eta = \begin{cases} s, s^*, s^z & \text{for} \quad |q| < Q, \quad \{s, s^*, s^z\} \equiv \mathfrak{S} \\ S, S^*, S^z & \text{for} \quad |q| \geqq Q, \quad \{S, S^*, S^z\} \equiv \mathfrak{B} \end{cases} \tag{7b.10}$$

where Q is a wavevector cut-off to be fixed later. We now have to write the quasiprobability P and the Liouvillian as

$$\begin{aligned} W(\xi, \xi^*, \eta, t) &= W(s, s^*, s^z, S, S^*, S^z, t) \\ L(\xi, \xi^*, \eta) &= L_{\mathfrak{S}}(s, s^*, s^z) + L_{\mathfrak{B}}(S, S^*, S^z) + L_{\mathfrak{S}\mathfrak{B}}(s, s, s^z, S\,S^*, S^z). \end{aligned} \tag{7b.11}$$

The three parts of the Liouvillian read

$$L_{\mathfrak{S}} = \begin{Bmatrix} \text{same structure as } L(\xi, \xi^*, \eta);\ \xi \to s,\ \xi^* \to s^*,\ \eta \to s^z; \\ \boldsymbol{q}\text{-summations restricted as } 0 \leqq |\boldsymbol{q}| < Q \text{ for all } s, s^*, s^z \end{Bmatrix}$$

$$L_{\mathfrak{B}} = \begin{Bmatrix} \text{same structure as } L(\xi, \xi^*, \eta);\ \xi \to S,\ \xi^* \to S^*,\ \eta \to S^z; \\ \boldsymbol{q}\text{-summations restricted as } Q \leqq |\boldsymbol{q}| \lesssim 1/a \text{ for all } S, S^*, S^z \end{Bmatrix}$$

$$L_{\mathfrak{S}\mathfrak{B}} = \sum_{\nu=1}^{\infty} \sum_{\mu=0}^{\min(2,\nu)} L^{(\nu,\mu)} \tag{7b.12}$$

$$L^{(\nu,\mu)} = \begin{Bmatrix} \text{same structure as } L(\xi, \xi^*, \eta); \text{ in each term there are} \\ \nu \text{ low-}|\boldsymbol{q}| \text{ variables } \mu \text{ of which occur as factors and} \\ (\nu-\mu) \text{ in derivatives whereas all other spin variables} \\ \text{are high-}|q| \text{ ones; } \boldsymbol{q}\text{-summations restricted as } 0 \leqq |\boldsymbol{q}| < Q \\ \text{for each low-}|\boldsymbol{q}| \text{ variable and as } Q \leqq |\boldsymbol{q}| \lesssim 1/a \text{ for each} \\ \text{high-}|\boldsymbol{q}| \text{ variable.} \end{Bmatrix}$$

Somewhat loosely speaking we may call $L_{\mathfrak{S}}$, $L_{\mathfrak{B}}$, and $L_{\mathfrak{S}\mathfrak{B}}$ the Liouvillians referring to the free motion of the long-wavelength variables ($L_{\mathfrak{S}}$), the short-wavelength variables ($L_{\mathfrak{B}}$), and their interaction, respectively. It turns out to be convenient for the following to further classify the terms in $L_{\mathfrak{S}}$, $L_{\mathfrak{B}}$, and $L_{\mathfrak{S}\mathfrak{B}}$ according to the number j of derivatives they contain

$$\begin{aligned} L_{\mathfrak{S}} &= \sum_{j=1}^{\infty} L_{\mathfrak{S}_j} \\ L_{\mathfrak{B}} &= \sum_{j=1}^{\infty} L_{\mathfrak{B}_j} \\ L^{(\nu,\mu)} &= \sum_{j=\max(1,\nu-\mu)}^{\infty} L_j^{(\nu,\mu)} . \end{aligned} \tag{7b.13}$$

δ) Orders of Magnitude and Time Scales

We assume that the spin system is in or nearly in thermal equilibrium at $T \gtrsim T_c$. Then we have as natural scales for the magnitude of spinfluctuations the thermal equilibrium expectation values

$$\sqrt{\langle \hat{S}^z_{\boldsymbol{q}} \hat{S}^z_{-\boldsymbol{q}} \rangle} \approx \sqrt{\langle \hat{S}^+_{\boldsymbol{q}} \hat{S}^-_{-\boldsymbol{q}} \rangle} \approx \sqrt{kT\chi(\boldsymbol{q})} . \tag{7b.14}$$

Here $\chi(\boldsymbol{q})$ is the wavevector-dependent static susceptibility which we assume known. We may now estimate the relative weights of the various terms in the Liouvillian by replacing the variables occurring as factors by

$\sqrt{kT\chi(\boldsymbol{q})}$ and derivatives by $1/\sqrt{kT\chi(\boldsymbol{q})}$. We thus obtain

$$
\begin{aligned}
O(L_{\mathfrak{S}_j}) &= \sum_{\boldsymbol{q}\boldsymbol{q}'\boldsymbol{q}_1\boldsymbol{q}_2\ldots\boldsymbol{q}_{j-1}}^{\text{all }|\boldsymbol{q}|<Q} A(\boldsymbol{q},\boldsymbol{q}',\boldsymbol{q}_1,\boldsymbol{q}_2,\ldots,\boldsymbol{q}_{j-1})\\
O(L_{\mathfrak{B}_j}) &= \sum_{\boldsymbol{q}\boldsymbol{q}'\boldsymbol{q}_1\boldsymbol{q}_2\ldots\boldsymbol{q}_{j-1}}^{\text{all }Q\leq|\boldsymbol{q}|\leq\boldsymbol{q}_{\max}} A(\ldots)\\
O(L_j^{(\nu,0)}) &= \sum_{\boldsymbol{q}\boldsymbol{q}'\boldsymbol{q}_{\nu+1}\boldsymbol{q}_{\nu+2}\ldots\boldsymbol{q}_{j-1}}^{\text{all }Q\leq|\boldsymbol{q}|\leq\boldsymbol{q}_{\max}}\ \sum_{\boldsymbol{q}_1\boldsymbol{q}_2\ldots\boldsymbol{q}_\nu}^{\text{all }|\boldsymbol{q}|<Q} A(\ldots)\\
O(L_j^{(\nu,1)}) &= \sum_{\boldsymbol{q}\boldsymbol{q}_\nu\boldsymbol{q}_{\nu+1}\ldots\boldsymbol{q}_{j-1}}^{\text{all }Q\leq|\boldsymbol{q}|\leq\boldsymbol{q}_{\max}}\ \sum_{\boldsymbol{q}'\boldsymbol{q}_1\boldsymbol{q}_2\ldots\boldsymbol{q}_{\nu-1}} A(\ldots)\\
O(L_j^{(\nu,2)}) &= \sum_{\boldsymbol{q}_{\nu-1}\boldsymbol{q}_\nu\ldots\boldsymbol{q}_{j-1}}^{\text{all }Q\leq|\boldsymbol{q}|\leq\boldsymbol{q}_{\max}}\ \sum_{\boldsymbol{q}\boldsymbol{q}'\boldsymbol{q}_1\boldsymbol{q}_2\ldots\boldsymbol{q}_{\nu-2}} A(\ldots)\\
A(\ldots) &= \sqrt{1/N^j}\sqrt{1/kT}^{\,j-2}\,|J(\boldsymbol{q},\boldsymbol{q}')|\\
&\quad\cdot\sqrt{\frac{\chi(\boldsymbol{q})\,\chi(\boldsymbol{q}')}{\chi(\boldsymbol{q}_1)\,\chi(\boldsymbol{q}_2)\ldots\chi(\boldsymbol{q}_{j-1})\,\chi(\boldsymbol{q}-\boldsymbol{q}'-\boldsymbol{q}_1-\boldsymbol{q}_2\ldots-\boldsymbol{q}_{j-1})}}\,.
\end{aligned}
\tag{7b.15}
$$

These expressions can be evaluated once the exchange integral $J(\boldsymbol{q},\boldsymbol{q}')$ and the static susceptibility are known. Since a rough estimate will serve our purpose we use simple choices for these quantities. Molecular field theory gives for $\chi(q)$ at $T=T_c$, up to numerical factors of order unity,

$$
\chi(\boldsymbol{q}) = \frac{N}{\hbar J a^2 \boldsymbol{q}^2}\,,
\tag{7b.16}
$$

where $\hbar J$ is the exchange energy for nearest neighbors in the lattice. The spherical continuum model for the lattice [91] gives for $J(\boldsymbol{q},\boldsymbol{q}')$

$$
J(\boldsymbol{q},\boldsymbol{q}') = J a^2\left(\frac{\boldsymbol{q}'^2}{1+a^2\boldsymbol{q}'^2} - \frac{\boldsymbol{q}^2}{1+a^2\boldsymbol{q}^2}\right).
\tag{7b.17}
$$

By replacing $\boldsymbol{q}$-sums in (7b.15) with integrals as

$$
\sum_{\boldsymbol{q}} \to V(2\pi)^{-3}\int d^3\boldsymbol{q} \approx N a^3 \int d^3\boldsymbol{q}
\tag{7b.18}
$$

and by suppressing numerical factors of order unity we obtain

$$
\begin{aligned}
&O(L_{\mathfrak{S}_{j+1}})/O(L_{\mathfrak{S}_j}) = (Q\,a)^4\sqrt{\hbar J/kT_c}\\
&O(L_{\mathfrak{B}_{j+1}})/O(L_{\mathfrak{B}_j}) = O(L_{j+1}^{(\nu,\mu)})/O(L_j^{(\nu,\mu)}) = \sqrt{\hbar J/kT_c}\\
&O(L_{\mathfrak{S}_j})/O(L_{\mathfrak{B}_j}) = (Qa)^{3+4j}\,, \qquad O(L_j^{(\nu,0)})/O(L_{\mathfrak{B}_j}) = (Qa)^{4\nu}\\
&O(L_j^{(\nu,1)})/O(L_{\mathfrak{B}_j}) = (Qa)^{4\nu-2}\,, \qquad O(L_j^{(\nu,2)})/O(L_{\mathfrak{B}_j}) = (Qa)^{4\nu-6}\,.
\end{aligned}
\tag{7b.19}
$$

We now choose the wavevector cut-off to obey

$$Qa \ll 1\,. \tag{7b.20}$$

The second parameter entering the order-of-magnitude relations (7b.19), $\sqrt{\hbar J/kT_c}$ is the ratio of the exchange energy between nearest neighbors in the lattice to the thermal energy at the Curie temperature. Molecular field theory [92] yields this parameter as

$$\hbar J/kT_c \approx [ZS(S+1)]^{-1} \tag{7b.21}$$

where Z is the number of nearest neighbors of a given spin in the lattice and S the total spin quantum number for the individual spins in the lattice. Resibois et al. [79, 80] assume this ratio to be small compared to unity. We won't have to require this.

We may now draw the following conclusions. Because of (7b.20) we can drop derivatives of higher than first order in the low-$|\boldsymbol{q}|$ Liouvillian $L_{\mathfrak{S}}$,

$$L_{\mathfrak{S}} = L_{\mathfrak{S}1}\,. \tag{7b.22}$$

This is not so for $L_{\mathfrak{B}}$ nor $L_{\mathfrak{S}\mathfrak{B}}$ unless $\sqrt{\hbar J/kT_c} \ll 1$ which we don't assume. The interaction Liouvillian does simplify, however, according to

$$L_{\mathfrak{S}\mathfrak{B}} = L^{(2,2)} + L^{(1,1)} + L^{(1,0)}\,. \tag{7b.23}$$

Moreover, we see that

$$O(L_{\mathfrak{S}}) \ll O(L_{\mathfrak{S}\mathfrak{B}}) \ll O(L_{\mathfrak{B}})\,. \tag{7b.24}$$

Since the Liouvillian has the dimension of an inverse time we may consider (7b.24) as an order-of-magnitude relation for the time scales characteristic for the processes described by $L_{\mathfrak{S}}$, $L_{\mathfrak{B}}$, and $L_{\mathfrak{S}\mathfrak{B}}$. The result (7b.24) thus justifies our separation of the set of spin variables in low-$|\boldsymbol{q}|$ and high-$|\boldsymbol{q}|$ subsets.

ε) Master Equation for the Reduced Quasiprobability over the Long-Wavelength Variables

Since the long-wavelength variables move on time scales much larger than the short-wavelength ones we can adiabatically eliminate the short-wavelength variables from (7b.8). The reduced quasiprobability over the low-variables,

$$\varrho(s, s^*, s^z, t) = \mathfrak{J}_{\mathfrak{B}}\, W(s, s^*, s^z, S, S^*, S^z, t)$$

with

$$\mathfrak{J}_{\mathfrak{B}} = \int \prod_{|\boldsymbol{q}| \geq Q} \mathrm{d}^2 S(\boldsymbol{q})\, \mathrm{d}^2 S^z(\boldsymbol{q}) \tag{7b.25}$$

obeys the formally exact Nakajima-Zwanzig equation (2b.13). The projector $\mathfrak{P}$ used to eliminate the high-$|\boldsymbol{q}|$ variables has to be taken as

$$\mathfrak{P} = B_{\text{ref}}(S, S^*, S^z)\,\mathfrak{I}_{\mathfrak{B}}\,. \tag{7b.26}$$

We choose the reference state for the high-$|\boldsymbol{q}|$ variables as the thermal equilibrium state

$$B_{\text{ref}}(S, S^*, S^z) = \mathfrak{I}_{\mathfrak{S}}\,\bar{W}(s, s^*, s^z, S, S^*, S^z)$$

with

$$\mathfrak{I}_{\mathfrak{S}} = \int\Bigl(\prod_{0 \leq |\boldsymbol{q}| < Q} \mathrm{d}^2 s(\boldsymbol{q})\Bigr)\Bigl(\mathrm{d}s^z(0) \prod_{0 < |\boldsymbol{q}| < Q} \mathrm{d}^2 s^z(\boldsymbol{q})\Bigr) \tag{7b.27}$$

and

$$\bar{W}(s, s^*, s^z, S, S^*, S^z) \leftrightarrow \hat{\bar{W}} = \mathrm{e}^{-\beta\hat{H}}/\mathrm{tr}\,\mathrm{e}^{-\beta\hat{H}}\,.$$

This is a reasonable reference state since the slowly moving long-wavelength variables see the rapidly moving short-wavelength variables as in thermal equilibrium. For the same reason a representative initial state for the whole system to start out with at some arbitrary time will be the local equilibrium state

$$W(s, s^*, s^z, S, S^*, S^z, t=0) = \varrho(s, s^*, s^z, t=0)\,B_{\text{ref}}(S, S^*, S^z)\,. \tag{7b.28}$$

This assigns an arbitrary initial distribution $\varrho(s, s^*, s^z, 0)$ to the low-$|\boldsymbol{q}|$ variables but takes the high-$|\boldsymbol{q}|$ variables as in thermal equilibrium. The Nakajima-Zwanzig equation for $\varrho(s, s^*, s^z, t)$ then describes the relaxation of the low-$|\boldsymbol{q}|$ variables to thermal equilibrium. Before writing down this equation explicitly let us state that the inhomogeneity $I(t)$ occurring there vanishes identically in the present case since we have

$$(1-\mathfrak{P})\,W(s, s^*, s^z, S, S^*, S^z, 0) = 0$$

because of (7b.26) and (7b.28). Furthermore, we have the identities

$$\begin{aligned} &\mathfrak{P} L_{\mathfrak{B}} = 0 \\ &\mathfrak{P} L_{\mathfrak{S}} = L_{\mathfrak{S}} \mathfrak{P} \\ &\mathfrak{P} L_{\mathfrak{S}\mathfrak{B}} \mathfrak{P} = 0\,. \end{aligned} \tag{7b.29}$$

The first of these holds since $\mathfrak{I}_{\mathfrak{B}} L_{\mathfrak{B}} X = 0$ by partial integration for reasonable X (such that surface integrals vanish). The second identity holds since the integration over the S^α commutes with differentiations with respect to the s^α. Finally, $\mathfrak{P} L_{\mathfrak{S}\mathfrak{B}} \mathfrak{P} = 0$ since all terms in $L_{\mathfrak{S}\mathfrak{B}}$ containing derivatives with respect to the S^α vanish by partial integration and all others by conservation of total spin and momentum. For instance,

$$\mathfrak{I}_{\mathfrak{B}} S^z(\boldsymbol{q})\,B_{\text{ref}} = \delta(\boldsymbol{q}, 0)\,\mathfrak{I}_{\mathfrak{B}} S^z(\boldsymbol{q})\,B_{\text{ref}} = 0\,,$$

since there is no high-$|\boldsymbol{q}|$ variable with $\boldsymbol{q}=0$. On the other hand,

$$\sum_{\boldsymbol{q}\boldsymbol{q}'} J(\boldsymbol{q}',\boldsymbol{q}'-\boldsymbol{q})\frac{\partial}{\partial s^{\alpha}(\boldsymbol{q})}\mathfrak{I}_{\mathfrak{B}}S^{\beta}(\boldsymbol{q}')\,S^{\gamma}(\boldsymbol{q}'-\boldsymbol{q})^{*}\,B_{\text{ref}}=0$$

because of $J(\boldsymbol{q},\boldsymbol{q})=0$. We thus find the Nakajima-Zwanzig equation (2b.13) to read here

$$\begin{aligned}\dot{\varrho}(s,s^{*},s^{z},t)=&-iL_{\mathfrak{S}}\varrho(s,s^{*},s^{z},t)-\int_0^t \mathrm{d}t'\,\mathfrak{I}_{\mathfrak{B}}L_{\mathfrak{S}\mathfrak{B}}\\ &\cdot \mathrm{e}^{-i(1-\mathfrak{P})Lt'}(L_{\mathfrak{S}\mathfrak{B}}+L_{\mathfrak{B}})\,B_{\text{ref}}\varrho(s,s^{*},s^{z},t-t')\,.\end{aligned}\tag{7b.30}$$

This is still exact. To within corrections of relative weight $(Q\,a)^2$ we may insert (7b.22) and (7b.23) and, by appealing to (7b.24), replace the exponential in the integral kernel with $\exp(-iL_{\mathfrak{B}}t)$, and neglect retardation effects. We thus obtain

$$\dot{\varrho}(s,s^{*},s^{z},t)=\Lambda\,\varrho(s,s^{*},s^{z},t)\tag{7b.31}$$

with the differential operator

$$\begin{aligned}\Lambda=&-iL_{\mathfrak{S}1}-\int_0^\infty \mathrm{d}t\,\mathfrak{I}_{\mathfrak{B}}(L^{(2,2)}+L^{(1,1)}+L^{(1,0)})\,\mathrm{e}^{-iL_{\mathfrak{B}}t}\\ &\cdot(L^{(2,2)}+L^{(1,1)}+L^{(1,0)})\,B_{\text{ref}}\\ =&-iL_{\mathfrak{S}1}-\int_0^\infty \mathrm{d}t\,\mathfrak{I}_{\mathfrak{B}}L^{(1,0)}\mathrm{e}^{-iL_{\mathfrak{B}}t}(L^{(1,1)}+L^{(1,0)})\,B_{\text{ref}}\,.\end{aligned}\tag{7b.32}$$

Here we have accounted for the fact that all terms in $L^{(2,2)}$ and $L^{(1,1)}$ contain derivatives $\partial/\partial S^{\alpha}(q)$ so that $\mathfrak{I}_{\mathfrak{B}}(L^{(2,2)}+L^{(1,1)})=0$. Moreover, by inserting the explicit expressions (7b.12) for the $L^{(\nu,\mu)}$ we see that $L^{(2,2)}$ does not contribute[8] at all to Λ which then turns out to be the following Fokker-Planck differential operator

$$\begin{aligned}\Lambda=&-iN^{-1/2}\sum_{\boldsymbol{q}\boldsymbol{q}'}J(\boldsymbol{q}',\boldsymbol{q}'-\boldsymbol{q})\\ &\left\{\frac{\partial}{\partial s^{z}(\boldsymbol{q})}s(\boldsymbol{q}')s^{*}(\boldsymbol{q}'-\boldsymbol{q})+2\frac{\partial}{\partial s^{*}(\boldsymbol{q})}s^{*}(\boldsymbol{q}')s^{z}(\boldsymbol{q}'-\boldsymbol{q})-2\frac{\partial}{\partial s(\boldsymbol{q})}s(\boldsymbol{q}')s^{z}(\boldsymbol{q}-\boldsymbol{q}')\right\}\\ &+\sum_{\boldsymbol{q}}\left\{(\gamma_{\parallel}(\boldsymbol{q})+i\Delta_{\parallel}(\boldsymbol{q}))\frac{\partial}{\partial s^{z}(\boldsymbol{q})}s^{z}(\boldsymbol{q})\right.\\ &+(\gamma_{\perp}(\boldsymbol{q})-i\Delta_{\perp}(\boldsymbol{q}))\frac{\partial}{\partial s^{*}(\boldsymbol{q})}s^{*}(\boldsymbol{q})\\ &+(\gamma_{\perp}(\boldsymbol{q})+i\Delta_{\perp}(\boldsymbol{q}))\frac{\partial}{\partial s(\boldsymbol{q})}s(\boldsymbol{q})\\ &+4\sum_{\boldsymbol{q}}\left\{D_{\parallel}(\boldsymbol{q})\frac{\partial^2}{\partial s^{z}(\boldsymbol{q})\,\partial s^{z}(-\boldsymbol{q})}+D_{\perp}(\boldsymbol{q})\frac{\partial^2}{\partial s^{*}(\boldsymbol{q})\,\partial s(\boldsymbol{q})}\right\}.\end{aligned}\tag{7b.33}$$

[8] Because of momentum conservation.

We here have, in the first bracket, the reversible nonlinear drift terms stemming from $L_{\mathfrak{S}1}$ which describe a mode-mode coupling. The remaining drift and diffusion terms are due to the dissipative influence of the thermal equilibrium high-$|\boldsymbol{q}|$ spin fluctuations on the low-$|\boldsymbol{q}|$ variables. They involve the complex damping constants and the diffusion constants.

$$\begin{aligned}
&\gamma_{\|}(\boldsymbol{q})+i\Delta_{\|}(\boldsymbol{q})=\gamma_{\|}(-\boldsymbol{q})-i\Delta_{\|}(-\boldsymbol{q})\\
&=2N^{-1}\sum_{\boldsymbol{q}'\boldsymbol{q}''} J(\boldsymbol{q}',\boldsymbol{q}'-\boldsymbol{q})\,J(\boldsymbol{q}'',\boldsymbol{q})\int_0^\infty \mathrm{d}t\,\mathfrak{I}_{\mathfrak{B}}\\
&\quad\cdot S(\boldsymbol{q}')\,S^*(\boldsymbol{q}'-\boldsymbol{q})\,\mathrm{e}^{-iL_{\mathfrak{B}}t}\left\{\frac{\partial}{\partial S(\boldsymbol{q}''+\boldsymbol{q})}S(\boldsymbol{q}'')-\frac{\partial}{\partial S^*(\boldsymbol{q}''-\boldsymbol{q})}S^*(\boldsymbol{q}'')\right\}B_{\mathrm{ref}}\\
&D_{\|}(\boldsymbol{q})=D_{\|}(-\boldsymbol{q})=D^*(\boldsymbol{q}) \qquad\qquad (7\mathrm{b}.34)\\
&\qquad=\mathrm{Re}\,N^{-1}\sum_{\boldsymbol{q}'\boldsymbol{q}''} J(\boldsymbol{q}',\boldsymbol{q}'-\boldsymbol{q})\,J(\boldsymbol{q}''+\boldsymbol{q},\boldsymbol{q}'')\int_0^\infty \mathrm{d}t\,\mathfrak{I}_{\mathfrak{B}}\\
&\qquad\quad\cdot S(\boldsymbol{q}')\,S^*(\boldsymbol{q}'-\boldsymbol{q})\,\mathrm{e}^{-iL_{\mathfrak{B}}t}\,S(\boldsymbol{q}'')\,S^*(\boldsymbol{q}''-\boldsymbol{q})\,B_{\mathrm{ref}}\,.
\end{aligned}$$

Similar expressions are obtained for $\gamma_\perp+i\Delta_\perp$, $D_\perp$. For $T>T_c$, i.e. for the paramagnetic state we have, by symmetry,

$$\gamma_\perp(\boldsymbol{q})=\gamma_{\|}(\boldsymbol{q})\,,\qquad \Delta_\perp(\boldsymbol{q})=\Delta_{\|}(\boldsymbol{q})\,,\qquad D_\perp(\boldsymbol{q})=D_{\|}(\boldsymbol{q})\,. \tag{7b.35}$$

The γ are damping constants, the Δ frequencies of periodic long-wavelength spin excitations. These latter must not be confused with the usual spin waves which occur below T_c as Goldstone modes tied up with a symmetry-breaking spontaneous magnetization. The Δ don't vanish above T_c. In order to find the relative weights of the various terms in the Fokker Planck "Liouvillian" Λ we again apply the scheme explained in part δ of this section

$$\begin{aligned}
\frac{O(\text{linear drift})}{O(\text{nonlinear drift})}&=\frac{O\left(\int_0^\infty \mathrm{d}t\,L^{(1,0)}\mathrm{e}^{-iL_{\mathfrak{B}}t}L^{(11)}\right)}{O(L_{\mathfrak{S}1})}=1/N(Qa)^4\\
\frac{O(\text{diffusion})}{O(\text{nonlinear drift})}&=\frac{O\left(\int_0^\infty \mathrm{d}t\,L^{(1,0)}\mathrm{e}^{-iL_{\mathfrak{B}}t}L^{(1,0)}\right)}{O(L_{\mathfrak{S}1})}=1/N(Qa)^2\,.
\end{aligned}\tag{7b.36}$$

Since (Qa) is small but finite we see that the weight of the linear drift and diffusion terms in Λ is asymptotically small in the thermodynamic limit. At the Curie point, where our estimates are valid, the behavior of the long-wavelength fluctuations is determined exclusively by the nonlinear mode-mode coupling terms. However, the mode-mode-coupling terms will loose their predominance for T away from the Curie temperature. This can be verified by using the temperature-dependent susceptibility $\chi(q)$ in the order-of-magnitude estimates carried out above. We therefore expect the Fokker Planck equation (7b.31) to be valid in an interpolative sense for $|(T-T_c)/T_c| \ll 1$.

Because of

$$J(\boldsymbol{q}',\boldsymbol{q}'-\boldsymbol{q}),\quad \gamma(\boldsymbol{q}),\quad \Delta(\boldsymbol{q}),\quad D(\boldsymbol{q})\to 0 \quad \text{for} \quad \boldsymbol{q}\to 0 \tag{7b.37}$$

the Fokker Planck equation $\dot{\varrho}=\Lambda\varrho$ implies as it must the conservation of the total spin which is represented by the $q=0$-variables $s^\alpha(0)$. This is obvious from the fact that Λ does not contain derivatives $\partial/\partial s^\alpha(0)$. The total-spin variables $s^\alpha(0)$ thus enter Λ as parameters only. Therefore the Fokker Planck equation does not determine the dependence of $\varrho(s,s^*,s^z)$ on the $s^\alpha(0)$. It is thus consistent with (7b.31) and probably necessary (to within the accuracy we are working with, $(Qa)^2$) to take

$$\varrho(s,s^*,s^z,t)=\bar{\varrho}_0(s(0),s^*(0),s^z(0))\,\tilde{\varrho}(s,s^*,s^z,t) \tag{7b.38}$$

where $\bar{\varrho}_0(s(0),s^*(0),s^z(0))$ is the equilibrium distribution of the total-spin variables and $\tilde{\varrho}(s,s^*,s^z,t)$ the time-dependent distribution of the $\boldsymbol{q}\neq 0$-spin variables. The latter obeys a Fokker Planck equation $\dot{\tilde{\varrho}}=\tilde{\Lambda}\tilde{\varrho}$ found by integrating (7b.31) over the $s^\alpha(0)$ and using (7b.38). The differential operator $\tilde{\Lambda}$ differs from Λ as given by (7b.33) in that the $\boldsymbol{q}$-sums exclude $\boldsymbol{q}=\boldsymbol{q}'$ and that spin-wave terms

$$i\sum_{\boldsymbol{q}}\omega(\boldsymbol{q})\left(\frac{\partial}{\partial s^*(\boldsymbol{q})}s^*(\boldsymbol{q})-\frac{\partial}{\partial s(\boldsymbol{q})}s(\boldsymbol{q})\right) \tag{7b.39}$$

appear where $\omega(\boldsymbol{q})$ are the wellknown spin-wave frequencies

$$\begin{aligned}\omega(\boldsymbol{q})&=2N^{-1/2}J(0,\boldsymbol{q})\int d^2s(0)\,ds^z(0)\cdot s^z(0)\,\bar{\varrho}_0(s(0),s^*(0),s^z(0))\\ &=2N^{-1/2}J(0,\boldsymbol{q})\langle\hat{S}_0^z\rangle\,.\end{aligned} \tag{7b.40}$$

It is easily checked that the Fokker Planck equation (7b.31) implies detailed balance [93, 94]. This has to be so since we have obtained (7b.31) by eliminating irrelevant variables from the time-reversal-invariant Liouville-von Neumann equation $\dot{\hat{W}}(t)=(-i/\hbar)\,[\hat{H},\hat{W}(t)]$. With the help of this important property of our Fokker Planck equation we easily find the stationary quasiprobability distribution to be a Gaussian with respect to the $q\neq 0$-spin variables. The width of the Gaussian turns out to be given by the moments

$$\begin{aligned}\mathfrak{I}_{\mathfrak{S}}\,s^z(\boldsymbol{q})\,s^z(-\boldsymbol{q})\,\bar{\varrho}(s,s^*,s^z)&=\langle\hat{S}^z_{\boldsymbol{q}}\hat{S}^z_{-\boldsymbol{q}}\rangle=D_{\|}(\boldsymbol{q})/\gamma_{\|}(\boldsymbol{q})\\ \mathfrak{I}_{\mathfrak{S}}\,s(\boldsymbol{q})\,s^*(\boldsymbol{q})\,\bar{\varrho}(s,s^*,s^z)&=\langle\hat{S}^-_{\boldsymbol{q}}\hat{S}^+_{-\boldsymbol{q}}\rangle=D_{\perp}(\boldsymbol{q})/\gamma_{\perp}(\boldsymbol{q})\,.\end{aligned} \tag{7b.41}$$

This relation between the diffusion and damping coefficients and equilibrium expectation values can be read as a dissipation-fluctuation theorem for the random process described by (7b.31).

The Fokker Planck equation (7b.31) is stochastically equivalent [27] to Kawasaki's Langevin equations. Its time-dependent solution and the corresponding spin correlation functions can be constructed by using a perturbation scheme developed by Kawasaki [81] and made systematic by Martin et al. [95]. We will not enter this problem here since our aim was to put the Fokker-Planck equation (7b.31) and the equivalent Langevin equations on a microscopic basis without recourse to any a-priori assumptions.

It is a pleasure to acknowledge helpful conversations with T. Arecchi, R. Bonifacio, V. Degiorgio, D. Forster, R. J. Glauber, R. Graham, H. Haken, B. Lix, P. C. Martin, N. E. Rehler, H. Risken, P. Schwendimann, M. O. Scully, and W. Weidlich.

References

1. Pauli, W.: Festschrift zum 60. Geburtstage A. Sommerfeld, S. 30. Leipzig: Hirzel 1928.
2. van Hove, L.: Physica **23**, 441 (1957).
3. Nakajima, S.: Progr. Theor. Phys. **20**, 948 (1958).
4. Zwanzig, R.: J. Chem. Phys. **33**, 1338 (1960) and Lect. Theoret. Phys. (Boulder) **3**, 106 (1960).
5. Montroll, E. W.: Fundamental problems in statistical mechanics, compiled by Cohen, E. G. D. Amsterdam: North Holland Publ. Co. 1962.
6. Prigogine, I., Resibois, P.: Physica **27**, 629 (1961).
7. Zwanzig, R.: Physica **30**, 1109 (1964).
8. Argyres, P. N., Kelley, P. L.: Phys. Rev. **134**, A98 (1964).
9. Wigner, E. P.: Phys. Rev. **40**, 749 (1932).
10. Moyal, J. E.: Proc. Cambridge Phil. Soc. **45**, 99 (1948); **45**, 545 (1949).
11. Glauber, R. J.: Phys. Rev. **130**, 2529 (1963); **131**, 2766 (1963).
12. Mehta, C. L., Sudarshan, E. C. G.: Phys. Rev. **138**, B274 (1965).
13. Cahill, K. E., Glauber, R. J.: Phys. Rev. **177**, 1857 (1969); **177**, 1882 (1969).
14. Agarwal, G. S., Wolf, E.: Phys. Rev. Letters **21**, 180 (1968).
15. Haken, H.: Z. Physik **219**, 411 (1969).
16. Haken, H., Weidlich, W.: Z. Physik **205**, 96 (1967).
17. Haake, F.: Phys. Rev. **A3**, 1723 (1971).
18. Bonifacio, R., Haake, F.: Z. Physik **200**, 526 (1967).
19. Lax, M.: Phys. Rev. **172**, 350 (1968).
20. Graham, R., Haake, F., Haken, H., Weidlich, W.: Z. Physik **213**, 21 (1968).
21. Gnutzmann, U.: Z. Physik **225**, 364 (1969).
22. Wangsness, R. K., Bloch, F.: Phys. Rev. **89**, 728 (1953).
23. Weidlich, W., Haake, F.: Z. Physik **185**, 30 (1965).
24. Senitzky, J. R.: Phys. Rev. **119**, 670 (1960); **124**, 642 (1961).
25. Mori, H.: Progr. Theor. Phys. **33**, 423 (1965).
26. Bonch Bruevich Tyablikov: The Green's function method in statistical mechanics. Amsterdam: North-Holland 1962.
27. Stratonivich, R. L.: Topics in the theory of random noise. New York: Gordon and Breach 1963.

28. Wang, N. Ch., Uhlenbeck, G. E.: Rev. Mod. Phys. **17**, 323 (1945).
29. Haake, F.: Z. Physik **223**, 364 (1969).
30. Weidlich, W., Haake, F.: Z. Physik **186**, 203 (1965).
31. Agarwal, G. S.: to be published.
32. Fröhlich, H.: Phys. Rev. **79**, 845 (1950).
33. Maxwell, E.: Phys. Rev. **78**, 477 (1950).
34. Reynolds, C. A. et al.: Phys. Rev. **78**, 487 (1950).
35. Cooper, L. N.: Phys. Rev. **104**, 1189 (1956).
36. Bardeen, J., Cooper, L. N., Schrieffer, I. R.: Phys. Rev. **108**, 1175 (1957).
37. Meservey, R., Schwartz, B. B.: In: Superconductivity, Kap. 4, edited by Parks, R. D. New York: Marcel Dekker, Inc. 1969.
38. Ginsberg, D. M., Hebel, L. C.: In: Superconductivity, Kap. 4, edited by Parks, R. D. New York: Marcel Dekker, Inc. 1969.
39. Scalapino, D. L.: In: Superconductivity, Kap. 10, edited by Parks, R. D. New York: Marcel Dekker, Inc. 1969.
40. Eliashberg, G. M.: Zh. Eksperim. i. Teor. Fiz. **38**, 966 (1960); Sov. Phys. JETP **11**, 696 (1960).
41. Scalapino, D. J., Schrieffer, I. R., Wilkins, J. W.: Phys. Rev. **148**, 263 (1966).
42. Yang, C. N.: Rev. Mod. Phys. **34**, 694 (1962).
43. Schwinger, J., Martin, P. C.: Phys. Rev. **115**, 1342 (1959).
44. Migdal, A. B.: Zh. Eksperim. Teor. Fiz. **34**, 139 (1958); Sov. Phys. JETP **7**, 996 (1958).
45. Hohenberg, P. C.: Proceedings of the Conference on Fluctuations in Superconductors, Asilomar, 1968, ed. by W. S. Goree and F. Chilton, Low Temp. Phys. Dept. Stanford Res. Inst. Menlo Park, California 94025.
46. Bogolyubov, N. N.: Physica **26**, 1 (1960).
47. Dicke, R. M.: Phys. Rev. **93**, 493 (1954).
48. Ernst, V., Stehle, P.: Phys. Rev. **176**, 1456 (1968).
49. Dialetis, D.: Phys. Rev. **A 2**, 599 (1970).
50. Lehmberg, R. H.: Phys. Rev. **A 2**, 833 (1970).
51. Agarwal, G. S.: Phys. Rev. **A 2**, 2038 (1970).
52. Rehler, N. W., Eberly, J. H.: Phys. Rev. **A 3**, 1735 (1971).
53. Bonifacio, R., Schwendimann, P., Haake, F.: Phys. Rev. **A 4**, 302 (1971).
54. Bonifacio, R., Schwendimann, P., Haake, F.: Phys. Rev. **A 4**, 854 (1971).
55. Degiorgio, V.: Optics Com. **2**, 362 (1971).
56. Bonifacio, R., Gronchi, M.: Nuovo Cim. Letters **I**, 1105 (1971).
57. Degiorgio, V., Ghielmetti: Phys. Rev. A4, 2415 (1971).
58. Haake, F., Glauber, R.: Phys. Rev. A**5**, 1457 (1972).
59. Bonifacio, R., Banfi, G.: to be published.
60. Schwendimann, P.: to be published.
61. Haken, H.: Handbuch der Physik, Bd. XXV/2c. Berlin: Springer 1970.
62. Kryukov, P. G., Letokhov, V. S.: Sov. Phys. **12**, 641 (1970).
63. Lamb, G. L., Jr.: Rev. Mod. Phys. **43**, 99 (1971).
64. Arecchi, T., Courtens, E.: Phys. Rev. **A 2**, 1730 (1970).
65. Scharf, G.: Helv. Phys. Acta **43**, 806 (1970).
66. Bonifacio, R., Preparate, G.: Phys. Rev. **A 2**, 336 (1970).
67. Glauber, R. I.: In: Quantum optics and electronics. New York: Gordon and Breach 1965.
68. Arecchi, T., Degiorgio, V.: Phys. Rev. **A 3**, 1108 (1970).
69. Lax, M., Louisell, W. H.: IEEE J. Quant. Electr. QE **3**, 47 (1967).
70. Risken, H.: Z. Physik **186**, 85 (1965); Fortschr. Physik **16**, 261 (1968).
71. Risken, H., Vollmer, H.: Z. Physik **201**, 323 (1967); **204**, 240 (1967).
72. Hempstead, R. D., Lax, M.: Phys. Rev. **161**, 350 (1967).

73. van der Pol, B.: Phil. Mag. **3**, 65 (1927).
74. Arecchi, T. et al.: Phys. Rev. Letters **16**, 32 (1966); Phys. Letters **25A**, 59 (1967).
75. Meltzer, D., Mandel, L.: Phys. Rev. **A3**, 1763 (1971).
76. Gerhardt, H., Welling, H., Güttner, A.: Z. Physik **253**, 113 (1972).
77. Gnutzmann, U.: Z. Physik **222**, 283 (1969).
78. Haken, H.: Z. Physik **181**, 96 (1964); **182**, 346 (1965).
79. Resibois, P., de Leener, M.: Phys. Rev. **178**, 806 (1969); 178, 819 (1969).
80. Resibois, P., Dewel, G.: Ann. Phys. (N. Y.) **69**, 299 (1972).
81. Kawasaki, K.: Ann. Phys. (N. Y.) **61**, 1 (1970); hier sind weitere einschlägige Arbeiten von Kawasaki zitiert.
82. van Hove, L.: Phys. Rev. **95**, 1374 (1954).
83. Landau, L. D., Khalatnikov, I. M.: Dokl. Akad. Nauk SSSR **90**, 469 (1954).
84. Halperin, B. I., Hohenberg, P. C.: Phys. Rev. **177**, 952 (1969).
85. Kadanoff, L. P.: Physics **2**, 263 (1966).
86. Kadanoff, L. P. et al.: Rev. Mod. Phys. **39**, 395 (1967).
87. Kadanoff, L. P., Swift, J.: Phys. Rev. **166**, 89 (1968).
88. Wyld, H. W., Jr.: Ann. Phys. (N. Y.) **14**, 143 (1961).
89. Lee, L. L.: Ann. Phys. (N. Y.) **32**, 292 (1965).
90. Haake, F.: Z. Physik **252**, 118 (1972).
91. Bennett, H. S., Martin, P. C.: Phys. Rev. **138**, A 608 (1965).
92. Kittel, C.: Introduction to Solid State Physics, Kap. 15. New York: Wiley 1968.
93. Graham, R., Haken, H.: Z. Physik **243**, 289 (1971).
94. Graham, R.: Springer Tracts in Modern Physics **66**, 1 (1973).
95. Martin, P. C.: private communication.

Dr. Fritz Haake
Institut für theoretische Physik
der Universität Stuttgart
D-7000 Stuttgart
Herdweg 77
Federal Republic of Germany

SPRINGER TRACTS IN MODERN PHYSICS

Ergebnisse der exakten Naturwissenschaften

Atomic Physics

Dettmann, K.: High Energy Treatment of Atomic Collisions (Vol. 58)
Donner, W., Süßmann, G.: Paramagnetische Felder am Kernort (Vol. 37)
Racah, G.: Group Theory and Spectroscopy (Vol. 37)
Seiwert, R.: Unelastische Stöße zwischen angeregten und unangeregten Atomen (Vol. 47)
Zu Putlitz, G.: Determination of Nuclear Moments with Optical Double Resonance (Vol. 37)

Elementary Particle Physics

Current Algebra

Furlan, G., Paver, N., Verzegnassi, C.: Low Energy Theorems and Photo- and Electroproduction Near Threshold by Current Algebra (Vol. 62)
Gatto, R.: Cabibbo Angle and $SU_2 \times SU_2$ Breaking (Vol. 53)
Genz, H.: Local Properties of σ-Terms: A Review (Vol. 61)
Kleinert, H.: Baryon Current Solving SU (3) Charge-Current Algebra (Vol. 49)
Leutwyler, H.: Current Algebra and Lightlike Charges (Vol. 50)
Mendes, R. V., Ne'eman, Y.: Representations of the Local Current Algebra. A Constructional Approach (Vol. 60)
Müller, V. F.: Introduction to the Lagrangian Method (Vol. 50)
Pietschmann, H.: Introduction to the Method of Current Algebra (Vol. 50)
Pilkuhn, H.: Coupling Constants from PCAC (Vol. 55)
Pilkuhn, H.: S-Matrix Formulation of Current Algebra (Vol. 50)
Renner, B.: Current Algebra and Weak Interactions (Vol. 52)
Renner, B.: On the Problem of the Sigma Terms in Meson-Baryon Scattering. Comments on Recent Literature (Vol. 61)
Soloviev, L. D.: Symmetries and Current Algebras for Electromagnetic Interactions (Vol. 46)
Stech, B.: Nonleptonic Decays and Mass Differences of Hadrons (Vol. 50)
Stichel, P.: Current Algebra in the Framework of General Quantum Field Theory (Vol. 50)
Stichel, P.: Current Algebra and Renormalizable Field Theories (Vol. 50)
Stichel, P.: Introduction to Current Algebra (Vol. 50)
Verzegnassi, C.: Low Energy Photo and Electroproduction, Multipole Analysis by Current Algebra Commutators (Vol. 59)
Weinstein, M.: Chiral Symmetry. An Approach to the Study of the Strong Interactions (Vol. 60)

Electromagnetic Interactions

Deep Inelastic Lepton Scattering

Drees, J.: Deep Inelastic Electron-Nucleon Scattering (Vol. 60)
Landshoff, P. V.: Duality in Deep Inelastic Electroproduction (Vol. 62)
Llewellyn Smith, C. H.: Parton Models of Inelastic Lepton Scattering (Vol. 62)
Rittenberg, V.: Scaling in Deep Inelastic Scattering with Fixed Final States (Vol. 62)
Rubinstein, H. R.: Duality for Real and Virtual Photons (Vol. 62)
Rühl, W.: Application of Harmonic Analysis to Inelastic Electron-Proton Scattering (Vol. 57)

Experimental Techniques

Panofsky, W. K. H.: Experimental Techniques (Vol. 39)

Form Factors, Colliding Beam Experiments

Photo- and Electroproduction of Pions

Quantum Electrodynamics

Field Theory, Light Cone Singularities

Strong Interactions

Mesons and Baryons

Regge Pole Theory, Dual Models

Symmetries

Weak Interactions

Nuclear Physics

Baryon-Baryon-Scattering

Electron Scattering

Nuclear Moments

Nuclear Structure

Weak Interactions

Optics

Laser

Plasma Physics

Relativity and Astrophysics

Cosmology

Solid-State Physics

Crystals

Electrons in Crystals

Magnetism

Optical Properties of Crystals

Quantum Statistics

To Appear in Volume 67

Ferrara, S., Gatto, R., Grillo, A. F.: Conformal Algebra in Space-Time and Operator Product Expansion

To Appear in Forthcoming Volumes:

Schmid, D.: Nuclear Magnetic Double Resonance – Principles and Applications in Solid State Physics
Bäuerle, D.: Vibrational Absorption of Electron and Hydrogen Centers in Ionic Crystals
Behringer, J.: Factor Group Analysis Revisited and Unified
Überall, H.: Study of Nuclear Structure by Muon Capture
Levinger, J. S.: Two-Nucleon and Three-Nucleon Systems
Brandmüller, J., Claus, R.: Light Scattering on Optical Phonons and Polaritons
Langbein, D.: Theory of van der Waals Attraction